I0505728

By faith Enoch was taken up so that he should not see death, and he was not found, because God had snatched him away. Now before he was taken, he constantly said, "God is pleased with me." Hebrews 11:5

WORMHOLES

From

Heaven

How We Get From Here To There

A Sunday School Teacher Takes A Look

GERALD McCRAY

WORMHOLES
From Heaven
HOW WE GET FROM HERE TO THERE

Front Cover Image courtesy of Benjamin Swanson
Ben@WaterwayAdvocates.org

Baby Faith Publishing

ISBN: 8614797829
ISBN-13: 9798614797829

WORMHOLES FROM HEAVEN

DEDICATION

To sci-fi fans and Bible students everywhere.

Theoretical physics is a guessing game.
(An educated guessing game albeit)

Almighty Most High GOD does not play games.

According to Romans 1:20 HE makes it crystal clear to the honest person that HE IS.

.

A Quick Note

As you journey through this Bible Study you will notice mention of Jesus' crucifixion in different ways which may seem ambiguous or confusing. You will see me using "the cross", "Jesus' cross", and "*The* Cross." The first two, I trust, are obvious but *The* Cross not so much. When I use *The* Cross, I am referring to the entire redemption plan which JESUS accomplished on behalf of spiritually dead humanity. By the entire redemption plan I refer to JESUS:

* Living in human flesh for thirty-three years getting full credit for living as a human so that HE could eventually give us full credit for divinity. (Psalm 82:6/2 Peter 1:3-4)

* Not preaching, teaching or doing any miracles (John 2:11) until after HE had the Holy Spirit come upon HIM (Matthew 3:16) as a man operating under the Old Covenant. Otherwise, it would be unfair for HIM to expect us to follow in HIS steps. (Luke 4:14-19/John 14:12/1 Peter 2:21)

* Did not operate out of HIS divinity but as a man being led by the Holy Spirit. (Matthew 9:36, 14:4, 15:32/Mark 1:41, 6:34, 8:2/Hebrews 5:1-5)

* Died in our place on our cross as a criminal.
  (Romans 3:25, 5:6, 5:8, 5:10)

* Went to Hell as a guilty man cursed by hanging on a
  tree.   (Acts 2:24, 2:27/Galatians 3:13)

* Was exonerated by being raised from the dead.
  (Acts 2:24, 13:33-37/Hebrews 1:4-8)

* Raised humanity up justified with HIS resurrection
  as long as we put faith in HIS work results not ours.
  (Acts 13:39, Romans 3:24,5:1,5:9/1 Corinthians
  6:11/Galatians 3:24/Titus 3:7)

* Imputes/shares HIS righteousness with the ungodly
  the very moment they put faith in HIS blood having
  their sins judged at the foot of HIS cross depositing
  HIS eternal life into our dead human spirit core and
  instantly making us the children of Almighty Most
  High GOD...forever.
  (Romans 5:21/1 Corinthians 1:30/Colossians 1:25-
  27, 3:4/2 Peter 1:1)

For God was in Christ, restoring the world to himself, **no
longer counting men's sins against them** *but
blotting them out. This is the wonderful message he has given
us to tell others.*   2 Corinthians 5:19-20 TLB

From the Garden of Gethsemane to the crucifixion

and going to Hell, JESUS was taking our place in our punishment. That was the cup of judgment which HE requested be taken from HIM when HE prayed in Gethsemane. It broke HIS eternal connection with GOD so that ours never will be.

…Jesus…by the grace of God…tasted death for all.
                                                    Hebrews 2:9

Revelation 14:10 speaks of the cup of judgment and wrath which those who reject JESUS will have to drink. That's right. If we reject JESUS's redemption plan for us which included HIM drinking the cup of judgment, indignation, and wrath; then we will have to drink our own cup of judgment. The redemption plan is so much more than we have realized or preached. Let's learn more about it.

CONTENTS

NOTEWORTHY

First of all, let me say that my observations and assertions are quite simplistic. It will soon become obvious that I am not scientifically gifted. Why should that stop me? After studying the Bible for a few months when I first became a Christian, I told my mother at the age of sixteen, "GOD did not tell the light to stop. That means it is still growing." That was my way of saying that the universe was still expanding, and I concluded that from studying the Bible. Interesting.

Wormholes or space-time bridges are such an integral part of pop culture entertainment on motion picture as well as the various small screens of today. We are so inundated with scientific postulates, axioms, and theories in visual form whether computer graphics on the big screen or in virtual environments that we wonder if they are real. After all, we think, those ideas and concepts came from somewhere.

In retrospect, we recognize seeds of the mobile flip phone in our favorite space frontier captain's communicator. That communicator also doubled as a phaser. Are we ever going to see that option? Imagine going to the app store to download the phaser app. How cool would that be? Butt dialing would definitely not be the worst thing that could go wrong if you sat on that phone. The places I go to in my mind would blow yours. Anyway, we think the possibilities of the ideas we see unfolding on film being real and wonder what if, but those concepts are based on well-founded assumptions or theorems.

In the world of *theoretical physics* scientists utilize what they know to be true (established laws of creation and motion) to postulate or guess about what they don't know. To be clear, educated guessing based on irrefutable laws of nature is quite practical. Where the science community has failed society is when it "preaches" or propagates anti-GOD motivated theorems as if they have graduated from theory to law.

For example, Sir Isaac Newton's laws of motion and Johannes Kepler's laws of planetary motion are irrefutable because they have been observed to behave the same way every time. That is what caused those theories to graduate to laws. Interestingly, though, Newton's law of gravity (universal gravitation) is absolutely irrefutable, but scientists still do not know what gravity is or how it works.

They simply know that it does. They do not understand why everything doesn't simply explode or implode. Isn't it interesting that scientists can explain so much except that tiny "X" factor which could put us face to face with GOD? HE wants us face to face with HIM by faith after honest observation.

For instance, the earth is travelling along its solar orbit at about 67,000 miles per hour while rotating on its axis about 1,050 miles per hour and no one has fallen off yet. That we know of that is. There is, though, a sort of cold case regarding a man named Enoch who up and disappeared about 3039 B.C.E. Scriptural references tell us that he experienced a personal rapture or snatching away by Creator GOD.

Possibly because Enoch knew too much. Or maybe it was because he testified for up to 300 years, "GOD is pleased with me!" Sounds like much more than positive confession. A faith confession, I surmise, is a positive type of confession that is convinced something has already happened. Maybe his faith opened a wormhole or simply caused a dimensional shift and Enoch slipped through to the other side.

The way his "disappearance" is worded in scripture, it could lead one to theorize that other people were present when "GOD took him." (Genesis 5:24) It sounds like this: Enoch was and then he was not. As if the event happened while others were watching and maybe even speaking with him. Maybe it was the last time they heard him testify that GOD was pleased with him and then…poof! Anyway, back to people not falling off this spinning rock which is dashing through space at more than breakneck speeds.

Scientists explain the magnetic cores of planets and the

pull which objects have on each other, but they don't know what gravity is or how it works. Although scientists cannot tell us everything there is to know about gravity, they can tell us that it simply works.[1] Maybe GOD will tell us one day what gravity is and how it works. Maybe Almighty Most High GOD is the reason gravity works. Maybe, just maybe HE is gravity…or something HE said is.

For well over two centuries and nearly three decades a religion of godlessness [adamant hope that there is no Creator GOD with moral absolutes] known as the theory of evolution has been preached in the primary schools, universities, and entertainment as gospel truth or scientific law. The truth of the matter is that evolution has never met the observational requirements which scientific law demands.

Evolution has never graduated to a law, and it never will because it is not science but a secular humanist belief system or rather hope system that there is no GOD. Why would anyone want that? If there is no GOD, then there are no moral absolutes. If there is no GOD, then there is no day of reckoning. If there is no GOD, then there is no judgment day. If there is no GOD, then humanity is the pinnacle of "creation" by some comic's blind watchmaker.

Uh sorry. I said creation again. If there is no GOD, then there is no Heaven to gain and no Hell to shun.

If there is no GOD, then when we are dead we are just done. If there is no GOD, then let's have as much perverted fun and debauchery in our years on earth as we can because…no GOD. Everything made by human hands is evidence of a designer. We fools refuse to use that honest deduction when we view nature and the cosmos. Are we that stupid or just dishonest?

…avoiding profane and vain babblings, and oppositions of **science falsely so called**: *By professing it, some people have deviated from the faith.*   Timothy 6:20-21

By following after and spouting the tenets of fake science as well as false doctrine some believers have even deviated from the faith. "The faith" is a plumb line or standard marker. Just as the captain's sextant is a standard for celestial navigation, "the faith" is a marker of established truths which are scripturally irrefutable. The love of GOD exhibited in the redemption plan is an established truth. The integrity and holy nature of GOD are established facets of HIS character. GOD will always be GOD and never change and evolution will always be a theory. And never graduate.

I find it quite telling that a community of the most educated and innovative thinkers is corrupted by a group which hates GOD so much that they will ignore empirical evidence of a deliberate designer and misrepresent findings which continue to point to HIS existence. Some scientists are hell bent on discovering anything which can prove to be a death knell to the existence of Creator GOD. The deeper some delve into their respective "disciplines" or rabbit holes, the more hope increases that they will ring that bell soon.

The space-time continuum and all of the exciting theoretical nuances which its postulates might produce, no doubt, cause some to salivate with anticipation. Wormholes, for example, might be the clapper which strikes that bell. Sadly, though, (for those who hold that false hope) the Bible is replete with examples of cross-spatial events as well as apparent time travel occurrences. Let's look at a brief introduction to wormholes from some notable sources. Then we will explore the Bible for some wow.

From www.space.com:

What Is Wormhole Theory?
By Nola Taylor Redd October 21, 2017

Wormhole Theory postulates that a theoretical passage through space-time could create shortcuts for long journeys across the universe.

…first theorized in 1916 by Austrian physicist Ludwig Flamm as a "white hole," a theoretical time reversal of a black hole.

In 1935, Albert Einstein and physicist Nathan Rosen

used the theory of general relativity to…propose the existence of "bridges" through space-time. These bridges would connect two different points in space-time, theoretically creating a shortcut that could reduce travel time and distance. The shortcuts came to be called Einstein-Rosen bridges, or wormholes.

From FoxNews.com:

Space 'Rain' And Black Holes Solve Galactic Star-Making Mystery
By Walt Bonner, | Fox News

For decades, the reason why some galaxies produce more stars than others has remained a mystery. The Milky Way, for instance, produces one new star a year while the Baby Boom Galaxy produces 4,000 per year, or one star every 2.2 hours. On the opposite end of the spectrum, some galaxies are barren, producing little to no stars anymore.

…galactic "rain" in the form of cool gas clouds which mist the cosmos, and their interaction with black holes. "*It's like we all of a sudden got really good at predicting the 'weather' in galaxies,*" Team Leader and MSU Professor Mark Voit told FoxNews.com.

Is there such a thing as I am proposing? I don't know

for sure but there are some recorded incidents in the Word of GOD which warrant a closer look.

Let's see what we find.

Wormholes From Heaven

Give praise to the Lord, O my soul.
O Lord my God, YOU are very
great; YOU are robed with honour
and power. YOU are clothed with
light as with a robe

Psalms 104:1-2a BBE

CHAPTER ONE

LIGHT BE!
OR
ALEPH BET?

Then God said, ["Light be!"], and there was light. And God saw the light, that it was good; and God divided the light from the darkness.

Genesis 1:3-4

During my Bible study, I have been absolutely fascinated by the even superficial acquaintance I have with the Hebrew alphabet. The first letter of the Hebrew alphabet is "aleph", and the second letter is "bet." What I find extremely intriguing is that each letter is not only a letter but a word. In the English language, an "a" is nothing more but an "a" but in Hebrew each letter is a 3-letter word. That may not be particularly interesting to you but when you spell out each letter, you end up with multiple words. Theoretically, you could spell particular letter words out indefinitely. One Rabbi Professor, Mordecai Kraft, explained how the chemical language most closely resembles the Hebrew language than any other language. And that "all languages come from the Hebrew." Fascinating.

The apparent "power" and infinite layers and levels of learning in Hebrew language study [or rather my superficial understanding of it is what ignited the title of this first chapter. IF the moment Almighty Most High GOD said, "LIGHT BE!" as recorded in the opening chapter of the book of Genesis is the moment when the theoretical "big bang" occurred; then I can only imagine HIM actually saying, "ALEPH BET!" That assumption is based on listening to several hours of lectures on the Hebrew language by at least 8 Rabbis, professors, language instructors, and the like. We know from Saul of Tarsus' Damascus Road conversion experience that the resurrected and ascended Jesus spoke Hebrew in that moment as He had, no doubt, in his discussion with Nicodemus. Is that an indication that Hebrew is the language of creation? Or at least the language of spiritual re-creation? Is Hebrew what the cosmos heard when Almighty Most High GOD said, "Big Bang," 138 billion years ago? Just joking unless "big bang" was translated into Hebrew. Another joke but I am not that good at it. Why 138 billion years? Search out astrophysicist Hugh Ross (reasons.org) who I first met when I purchased his book *The Fingerprint Of GOD* several years ago. Christians who are easily swayed by spiritually dead scientists will be genuinely blessed by Ross' writings and spirit. Anyway, scientists have used their wealth of knowledge to accurately determine the moment after the big bang occurred.

When YHWH said, "Light be," is it possible that the light came straight out of HIM? The Word does refer to HIM as light in several passages. And HIS Son as The Light as well.

This is the message which we have heard from Him and declare to you, that **God is light** *and in Him is no darkness at all. If we say that we have fellowship with Him, and walk in darkness, we lie and do not practice the truth. But if we walk in the light as He is in the light, we have fellowship with one another, and the blood of Jesus Christ His Son cleanses us from all sin.* 1 John 1:5-7

Scottish physicist Peter Higgs described the elementary particle known as the boson as tangible proof of an invisible and deliberate Creator. I am a big fan of the situational comedy **Big Bang Theory**. I did not begin watching it until after it entered syndication. One of the reasons I began watching it is the success my youngest daughter had in her High School physics classes. Interestingly, as I have begun listening to Rabbis and Jewish professors lecture on the nuances of the Hebrew alphabet, I have become increasingly interested in physics.

The power which is inherent in the first two letters of the Hebrew alphabet alone has led me to asking this question about the first statement on day one of creation. It was after watching at least 50 hours of lectures on the Hebrew language that I began imagining GOD's words of day one being, "Aleph Bet!" GOD knows. There is also the issue of the water which was present before the first command. Water was present before the first creation command, water is present in the natural realm, and water is present in the unseen realm according to several scriptural glimpses behind the curtain of nature. I could not help but wonder if water is a wormhole of sorts.

Sonoluminescence is the name of a process or experiment which introduces sound waves on the surface of water at the perfect frequency to generate light. There seems to be water everywhere in the new kingdom along with angelic beings incessantly shouting praises to Almighty Most High GOD. The correct sound frequency, sound volume, and water can result in something fantastic.

Look at John's glimpse into the future arrival of the Jerusalem City from Heaven coming to Earth to merge with the terrestrial Jerusalem. Praying for the peace or shalom of Jerusalem is not merely referring to the cessation of conflict and the eradication of the enemies of Almighty Most High GOD's chosen people. The peace of Jerusalem is the oneness or wholeness of Jerusalem. Praying for the peace of Jerusalem is praying for Jerusalem's healing. When the Heavenly Jerusalem merges with the terrestrial one, the prayer will finally be manifested, and Jerusalem will be made whole. And it's going to take some special light to heal her.

And I saw no temple in the city, for its temple is the Lord God the Almighty and the Lamb. And the city has no need of sun or moon to shine on it, for the glory of God gives it light, and its lamp is the Lamb. By its light will the nations walk, and the kings of the earth will bring their glory into it, and its gates will never be shut by day—and there will be no night there. Revelation 21:22-25 ESV

The sun will finally lose its usefulness when it comes

to the shalom of the New Jerusalem. King David's eternal capitol will be illuminated by the very glorious presence of Almighty Most High GOD and the resurrected Lamb of GOD. The eternal glorious presence of the Holy Trinity will light the largest Ark of the Covenant that any human has ever seen. The dimensions of the city along with the details of what's inside the city are quite telling. There is not a temple in the city because **the city is the temple**. Wherever the presence of Almighty Most High GOD is found is the temple. According to 1 Corinthians chapters three and six the spirit core or every new creation believer is home to the Spirit presence of Almighty Most High GOD. Does that make you a temple? Me?

The Word of GOD informs the new creation believer that Holy Spirit dwells inside each and every one of them making each a temple of the Living GOD. The same YHWH who called light into existence and wears light as a kingly robe commanded HIS light to shine out of darkness and inside every new creation believer. Why would anyone have a penchant for darkness?

In the coming age, the glory light of HIS Spirit presence will once again roam the earth in the cool of the day inside all of HIS blood washed new creation believers as HE lights up New Jerusalem. To those who read the promises and predictions of GOD's Word even though every prediction has come true as scheduled and on schedule and still refuse to believe that the remaining predictions will come to fruition; I have one thing to say.

I am not [usually] a gambling man but…do you want to aleph bet?

For behold, darkness shall cover the earth, and thick darkness the peoples; but the LORD will arise upon you, and his glory will be seen upon you.

And nations shall come to your light, and kings to the brightness of your rising.   Isaiah 60:2-3 ESV

CHAPTER TWO

THE *DAY*
THE LIGHTS
DIED OUT

*And the LORD GOD commanded the man, saying, "Of
every tree of the garden you may freely eat; but of the tree of
the knowledge of good and evil you shall not eat, for **in the
day** that you eat of it you shall surely die."* (disconnect
from my light of life)

Genesis 2:16-17

"In the day you disobey [MY instructions], you will surely die," YHWH (The LORD) warned Adam. The light which GOD's Spirit put inside the human structure was there to last forever. That is why it took so long for people to die early on. One man lived 969 years before death caught up with him. In fact, it was one of the sons of Enoch. You know, the cold case? Enoch is probably the only person in history to outlive his own son who lives so close to 1000 years. Enoch never died according to the Word of GOD so I guess he has an upcoming appointment along with another rapture. Even the one responsible for introducing death into creation lived 930 years before death caught up with him. So, wasn't he supposed to die "in the day he disobeyed?"

O, but he did. Because a day with the LORD is one thousand years according to II Peter 3:8; as long as sinful man did not live beyond a thousand years of age, the warnings of GOD would ring true. The fall of humanity resulted in the *darkness of perdition* replacing the glory light of GOD inside the spirit core of every descendant of Adam and Eve. From that point on, everyone in their bloodline would be born spiritually dead. When the replacement Light (Christ) arrived on Earth to prepare a way to reinstall the light permanently; the darkness which was in every other person on the earth could not comprehend or "lay siege" to the replacement Light. (John 1:5)

Arise, shine; For your light has come! And the glory of the LORD is risen upon you. For behold, the darkness shall cover the earth, **And deep [gross] darkness the people***; But the LORD will arise over you, And His glory will be seen upon you.* Isaiah 60:1-2 NKJV

Hear and give ear; be not proud, for the LORD has spoken. Give glory to the LORD your God before he brings darkness, before your feet stumble on the twilight mountains, and while you look for light he turns it into gloom and makes it **deep darkness***.* Jeremiah 13:15-16 ESV

The closer Jesus got to HIS appointment with Calvary, the less HE spoke. One reason HE gave was shared in John 14:30. "The [devil] has nothing in Me." There is none of the nature of the prince of darkness in Jesus Christ and HE was going to ensure that the devil gained nothing to which he could attach to Jesus by being more vigilant about the words coming out of HIS mouth more than ever before. The words of our mouths possibly open wormholes inside us which either let GOD's Light flow through or Satan's darkness attach to and grow in us. Watch your mouth!

On The Day Of Pentecost, The Lights Came Back On.

This is the message which we have heard from [Jesus] *and*

31

*declare to you, that **God is light** and in Him is no darkness at all.*   1 John 1:5

Because there is "***no darkness in HIM at all***," what does it mean that HE resides inside our resurrected spirit core? Is our spirit core as dark as it was before Holy Spirit took up residence there or is it full of light as the Word says about Almighty Most High GOD? According to the following scripture, GOD not only brought the light which is in HIM, but HE also commanded it to shine inside us just as HE ordered it to shine on day one of creation. When He commanded the light to shine inside us, was that day one of a spiritual creation in us? Can you imagine that?

*For it is the **God** who commanded light to shine out of darkness, who **has shone in our hearts** to give the light of the knowledge of the glory of God in the face of Jesus Christ.*   2 Corinthians 4:6

Although Holy Spirit was highly active in the Old Testament and under the different covenant dispensations of Noah, Abraham, the Levitical system, David, and the prophets; HE only, according to scriptural indication, rested upon those with whom HE partnered. His Old Testament ministry appeared to be external almost exclusively. One possible reason for that was that the spirit core of even the most holy Old Testament saint was spiritually dead and the blood of sacrificed animals could only "cover" sins as a rehearsal for the eternal sacrifice which would come at Calvary and cleanse sins forever. Christ's blood would wash away the sin nature.

Do you remember Jesus' contrast between the "greatest prophet of the Old Testament" -John Baptist- and the bottom-of-the-totem-pole blood washed saint? In Matthew 11 and Luke 7, Jesus said that the least in the kingdom of grace is greater than the greatest prophet of the Old Testament. But why? How? The Old Testament heroes of the faith and prophets which the Holy Spirit did magnificently powerful miracles and wonders in partnership with could only partner with Holy Spirit. They would experience the Spirit presence of GOD coming upon them as a cloak or cape.

Every, every, every blood washed new creation believer has become the new residence of the Holy Spirit since the veil in the temple was torn from top to bottom when Jesus screamed from the cross, "Tetelestai!" Jesus declared that humanity's sin debt was "paid in full" and ABBA Father GOD raised Him from the dead so that we humans would know that statement (and everything Jesus said and did) was true. Because Jesus laid the groundwork for Holy Spirit to turn the lights back on inside the human spirit core, even the new creation saint which no one would choose to be on their team is greater than the greatest prophet of the Old Testament. In fact, it sounds like that are greater than all of the Old Testament prophets combined.

Can that be possible?

In the first chapter of the Bible, we see what happened when **THE ETERNAL** (traditional reference to GOD) began creation with the words of HIS mouth. The darkness and expansive void changed the moment the light of Almighty Most High GOD showed up. Once the light came on things started to change. We understand from **Einstein's equivalence equation** that light energy is inside every morsel of matter. Light and matter are the same thing. YHWH used HIS words to create light and HE used the light to create matter.

Once the light showed up, the words from GOD's mouth started designing the cosmos, the upper and lower atmospheres, the earth's mantle, and the oceans. From the light came the birds of the air, the land animals, and the marine life. From the light came the dust from which humanity was made. Once GOD breathed HIS Spirit into Adam, his soul started to live or connect or liaison with the Spirit of GOD. The light of GOD was inside the first couple, and they were able to fellowship with HIM and walk with HIM in "the cool of the day" because HIS nature inside them.

*Do you not know that if you present yourselves to anyone as obedient slaves, you are slaves of the one whom you obey, either of **sin, which leads to death**, or of **obedience, which leads to righteousness**?*

Romans 6:16 ESV

We choose whom we will serve. We will either remain in the darkness of the sin nature or we will opt for the light of Jesus' righteousness. Doing nothing allows us to keep what we have – eternal death. Rejecting the grace which Jesus' cross affords us allows us to keep what we have. Accepting GOD's gift of eternal life through the blood, cross, and resurrection of Jesus Christ affords us the opportunity to trade our spiritual death for HIS spiritual life. The enemy of human souls says to everyone every single day, "Let's make a deal," and we decide which door we will take. The one which is well lit or the one which is bathed in gross darkness. To the natural eye, though, they look completely opposite.

But now that you have been set free from sin and have become slaves of God, the fruit you get leads to sanctification and its end, eternal life.   Romans 6:22 ESV

Once Adam and Eve **believed and acted on the words** of the devil/serpent in rebellion to the instructions of Almighty Most High GOD; they became slaves to the sin nature. They became children of darkness and damnation and every member of the human race (Acts 17:26), because we are in their bloodline, was born spiritually dead. To be spiritually dead means to be disconnected from the life and light of the Spirit of GOD. Everyone who came through Adam was born in darkness because Adam's rebellion resulted in the sin nature disconnecting humanity from the light of GOD.

Look at the result of rejecting GOD's light:

Then Jesus cried out and said, "He who believes in Me, believes not in Me but in Him who sent Me. **45** *And he who sees Me sees Him who sent Me.* **46 I have come as a light into the world,** *that whoever believes in Me should not abide in darkness.* **47** *And if anyone hears My words and does not believe, I do not judge him; for I did not come to judge the world but to save the world.* **48** *He who rejects Me, and does not receive My words, has that which judges him — the word that I have spoken will judge him in the last day.* **49** *For I have not spoken on My own authority; but the Father who sent Me gave Me a command, what I should say and what I should speak.* **50** *And I know that His command is everlasting life. Therefore, whatever I speak, just as the Father has told Me, so I speak."* John 12:44-50

Yeshua Jesus came to light up our spirit cores again.

HIS blood would resurrect the dead human spirit for anyone who would put faith in HIS blood and cross as full payment for the sin debt which they inherited from Adam. Faith in Jesus' resurrection for complete (Romans 10:9-10) and eternal justification (Hebrews 9:12-15) puts the formerly wicked ungodly soul instantly into Jesus' bloodline. We are HIS family now and enjoying everlasting redemption. We've been bought back (ransomed/redeemed) never to be lost again. Repurchased by the blood which never loses its value, worth or power.

So, the Father has turned the lights back on inside anyone who made the right deal and chose the blood of Jesus for the basis of their status with GOD. Just as Adam condemned all to eternal darkness and spiritual death; Jesus secured eternal glory light for anyone who will ask for it. That's right. Just ask.

Most assuredly, I say to you, he who hears My word and believes in Him who sent Me has everlasting life, and shall not come into judgment, but has passed from death into life. **25** *Most assuredly, I say to you, the hour is coming, and now is, when the dead will hear the voice of the Son of God; and those who hear will live.* **26** *For as the Father has life in Himself, so He has granted the Son to have life in Himself,* **27** *and has given Him authority to execute judgment also, because He is the Son of Man.* **28** *Do not marvel at this; for the hour is coming in which all who are* *in the graves will hear His voice **29** and come forth—those who have done good, to **the resurrection of life**, and those who have done evil, to **the resurrection of condemnation**. **30** I can of Myself do nothing. As I hear, I judge; and My judgment is righteous, because I do not seek My own will but the will of the Father who sent Me.*   John 5:24-30 NKJV

A resurrection of life and a resurrection of judgment? When was the last time you heard someone say that there will be at least two resurrections?   Church has not done a good job of teaching the Word of GOD the way it should have been taught – like a class instead of some arm flailing, mouth foaming tirade.   Preachers deliberately speaking in a manner which resembles symptoms of some sort of respiratory ailment has been traditional theater in many afro-centric houses of theater…uh worship… for ages.   But why?

GOD knows and so does the devil.   We should have been taught rather than accidentally spit on. We should have had deep foundations of spiritual and scriptural knowledge laid in us.   The right teaching and yes even preaching can make the hearer more "fall proof" than the wrong type of preaching and teaching.   Anyway…

Do you realize that if you have had your blood bath at the foot of the cross that Jesus Christ has already called you out of the grave of the sin nature (Romans 6,

Ephesians 2:1-5, Colossians 2:13) and raised you to eternal life right now (John 3:18, 5:24, Ephesians 2:6-8, Colossians 2:13-14).   That means that you are reserved a spot in the first resurrection along with inheritance gifts.   If you leave this world without having accepting GOD's unspeakable gift of grace, then you have reserved yourself a place in the second resurrection.   Don't make that deal.   There's no profit in it.   Only wages according to Romans 6:23.   Death.

Now we know that what things so ever the law says, it says to them who are under the law: that every mouth may be stopped, and all the world may become guilty before God.   Romans 3:19

CHAPTER THREE

WE WERE ALL BORN THIS GUILTYWAY

Behold, I was brought forth in iniquity, and in sin did my mother conceive me.
Psalm 51:5 ESV

The scientific and medical communities are baffled by the death of the human body. They can explain death but not why it exists or occurs to the human body. The fact that the human body counts down to an expiration date is a mystery. From scientific observation, the human body should function indefinitely. Even eternally. The human structure was originally designed to function into perpetuity. So, why do we die? Where did death come from? The fifth chapter of Romans tells us where death came from.

Sin came into the world through one man, and death through sin, and so death spread to all men...

Romans 5:12 ESV

Although natural death is the result of the sin nature, spiritual death is the main subject of the scripture above. Spiritual death is eternally more important than natural death. To be spiritually dead means to be disconnected from the Spirit of Life. Spiritual death came first, and physical death followed. Adam and Eve died spiritually the moment Adam ate in disobedient agreement with his wife. It took 930 years for physical death to catch up with him. When the light inside Adam and Eve turned to gross darkness; the darkness (spiritual death) was passed on to all of humanity just as the genetic traits of ancestors are inherited by their descendants. Spiritual genetics were handed down. Some really bad traits.

Sin came into the world through one man, and death through sin, and so death spread to all men because all sinned for sin indeed was in the world before the law was given, but sin is not counted where there is no law. Yet death reigned from Adam to Moses

Romans 5:12-14 ESV

The theme scripture for this chapter hearkens to the erroneous pop culture philosophy which assumes that because we were all born spiritually dead and separated from Almighty Most High GOD that we must accept ourselves the way we are. That train of though is a complete mistake. That mindset lends itself to an erroneous assumption that GOD accepts us the way we are. Well, sort of.

It is true to think that GOD accepts us the way we are, but it cannot stop there. The message of *The* Cross - Jesus' redemption mission—is that GOD loves us too much to leave us the way we are. We are born broken, dead in trespasses and sins, and eternally separated from the life of GOD. The darkness of evil and human brokenness is rampant. Within the first six months of a U.S. President's blackish governance hundreds of billions of taxpayer dollars are given to foreign entities to make pregnant wombs around the world less safe for the preborn than they were already. And proud of it apparently. He did his part to make more wombs into tombs. How spiritually dead can you get?

43

Ancient Egyptians and several other cultures were proud to burn their infants alive in demon cult religion practices worshiping the deity Molech. Molech had a bull's head and was usually seated with a fire burning in its hands, lap, and hollow core. Parents would place their first-born infants on Molech's red hot outreached hands and watch as their child burned alive. They believed that the sacrificing of the first born resulted in blessings for the rest of the family including long life and prosperity.

Is humanity born broken, spiritually dead, and guilty before the GOD of creation because of the sin nature? If the hell bent to make hundreds of billions of dollars available for offshore abortions is any indication of how broken the human soul is and how black the human heart can be then I guess that answers the question. Instead of making *killing the future* a priority of his first year in office, it may have been a better decision to invest those hundreds of millions of taxpayer dollars into the communities where the people are disadvantaged. I am sure the black community voted in traditional lock step for the political party which desperately devises funding schemes to kill as many preborn black babies as a 24-hour day allows. According to Grand Rapids Right To Life website (grrtl.org) over one thousand black babies alone are killed every single day in America. I guess those are the black lives which do not matter enough to protest over because no one is protesting the loss of those lives. Indifference to those black lives and any lives lost in the same way as these are cutoff in what has been called Planned Parenthood's Black Genocide is a definite clue to whether one is spiritually alive or not. Scriptural reference examples of Adam and Eve's descendants (the entire human race according to Acts:17:26) being born spiritually dead, broken, and in eternal need of the Savior Jesus Christ the Son of Creator GOD follow:

Genesis 2:7

Adam "***became a living soul***" means that he was on the same level as any wildlife until the Spirit of Life breathed spirit into his spirit core.

John 5:28-29

Those who die either spiritually dead (disconnected from GOD) or spiritually alive (saved, sanctified, Holy Spirit filled, and fire baptized) will be resurrected in that same state. The **first resurrection** of the living occurs at the rapture of the church when Christ collects HIS saints. Three and a half to seven years later THE LORD returns to rescue HIS chosen people, the Jewish Hebrew Israelites, from total destruction. The demonic leader of the One World Government who will convince the spiritually dead that he/she is **god over GOD** (G.O.G. -I know it's not this simple) will assemble the military might of most of the world to finally "wipe Israel off the face of the map" as many spiritually dead despots, dictators, and rulers, have vowed. The **second resurrection** (Revelation 20:5) occurs 1000 years after Jesus defeats Gog/MaGog and takes HIS prophesied throne of David in King David's eternal capitol – New Jerusalem. Praying for "the peace of Jerusalem" (Psalm 122:6) is praying for much more than the cessation of conflict but for the shalom (wholeness/oneness) of GOD's will and pattern in Heaven being accomplished on Earth. When the New Jerusalem descends from Heaven (Revelation 21:2-3) and merges with the war plagued earthly Jerusalem; the city of peace will finally be healed and whole.

Romans 5:10 ESV
When we were the spiritually dead enemies of GOD, Jesus died so that we could be reconciled to Almighty Most High One True Living GOD. Jesus became spiritually dead so that we could be made spiritually alive. That is amazing grace. Now, we only need to ask for it and live forever connected to the Spirit of Life.

Romans 5:12-19
Spiritual death came from Adam, and everyone born in his blood line were born spiritually dead. Agreeing to kneel at the cross of Christ and having our sins judged in Jesus on the cross results in us being made right with GOD based on what Jesus did and not what we do. Being born spiritually dead was the result of no effort of our own just as being made right with GOD is. Being born spiritually dead is the result of Adam's

works. Being reborn spiritually alive is the result of Jesus' work.

Romans 14:9
Jesus Christ dying spiritually and being resurrected to life again makes HIM the ruling Lord over the naturally and spiritually living as well as the naturally and spiritually dead.

Search out all of the moments in the book of the ***Acts Of*** [**The Holy Spirit Through**] ***The Apostles*** in which the Lord raised people from the dead working through HIS people. In fact, every act of the Spirit through the new creation believers was a great work. The seven sons of Sceva (Acts 19:11-20) attempted to replicate what they saw Paul and the Apostles doing but these sons of Sceva were not the spiritually alive children of Almighty Most High GOD. They thought Paul used the name of Jesus as a good luck charm or some "blessed" talisman used for exorcisms.

When they used the name of **the Jesus whom they did not know**, (excellent teaching by Pastor Dr. David Jeremiah) to cast out a demon; the demon spoke acknowledging Jesus and Paul (authority and delegated authority) but said that it did not recognize the authority of these men. The demon used the body of the man whom it possessed to beat some sense into Sceva's seven sons. The men ran home naked and wounded and when word of that incident got around several people of that town brought their black magic books, figurines, and trinkets to an impromptu bonfire to be burned up.

1 Corinthians 15:12-21
Because Christ got up from the dead then that guarantees the resurrection of everyone else; the spiritually living and the spiritually dead.

The epistle to the Gentile believers at Ephesus Speaks of the contrast between the before-the-cross condition or state of sin which is spiritual death with the eternal life which GOD has deposited into every new creation believer.

Ephesians 2:1
And you were dead in the trespasses and sins…

Ephesians 2:5
even when we were dead in our trespasses, made us alive together with Christ—by grace you have been saved

Colossians 2:13
And you, who were dead in your trespasses and the uncircumcision of your flesh, God made alive together with **Christ***, having forgiven us all our trespasses*

1 Thessalonians 1:10
and to wait for his Son from heaven, whom he raised from

the dead, Jesus who delivers us from the wrath to come.
I Thessalonians 4:16-18

For the Lord Himself will descend from heaven with a shout, with the voice of an archangel, and with the trumpet of God. And **the dead in Christ will rise first**. *Then we who are alive and remain shall be caught up* [raptured] *together with them in the clouds to meet the Lord in the air. And thus we shall always be with the Lord. Therefore* **comfort one another with these words**.

Encourage one another with the teachings of end time events and the return of the Lord Jesus Christ. The church needs to catch up to this imperative and teach eschatology, prophecy, and the end of the age events to give the people of Almighty Most High GOD peace when they see so much turmoil and distress in the new cycles.

James 2:26a
…The body apart from the spirit is dead….

1 Peter 1:3-4
Blessed be the God and Father of our Lord Jesus Christ, who according to His abundant mercy **has begotten us again to a living** [raised us from death to life] *hope through the resurrection of Jesus Christ from the dead, to an* **inheritance incorruptible** *and undefiled and that does not fade away,* **reserved in heaven for you.**

Being washed in the blood of the Lamb of GOD at the foot of the cross results in us being rescued from the plantation of the sin nature and called out of our graves (John 5:28-29) of spiritual death (Ephesians 2:1,5/Colossians 2:13) and raised to a newness of life (Romans 6:4) in the family (Hebrews 2:10) of Almighty Most High GOD so much so that we inherit HIS divine nature. (2 Peter 1:4)   Why would anyone want to remain in the same spiritually dead condition (state of sin) in which they were born?

Why would you?   How can we categorize those who will see the Heavenly Father face to face and still pray for death rather than mercy?   (Luke23:27-31/Revelation6:14-17)

"Almighty Most High GOD really does love me too much to leave me the way that I am, but I still do not want to accept HIS offer of grace, mercy, and forgiveness through Jewish Jesus Christ HIS Son," just might be the impetus and mindset here for rejecting GOD once you see HIM.   The old but incorrect adage "seeing is believing" is not necessary the case, I guess.

May grace and peace be multiplied to you in the knowledge of God and of Jesus our Lord.   3 His divine power has granted to us **all things that pertain to life and godliness***, through the knowledge of him who called us to his own glory and excellence,   4 by which he has granted to* *us his precious and very great promises, so* **that through them you may become partakers of the divine nature***, having escaped from the corruption that is in the world because of sinful desire.* **5** *For this very reason, make every effort to supplement your faith with virtue, and virtue with knowledge,* **6** *and knowledge with self-control, and self-control with steadfastness, and steadfastness with godliness,* **7** *and godliness with brotherly affection, and brotherly affection with love.* **8** *For if these qualities are yours and are increasing, they keep you from being ineffective or unfruitful in the knowledge of our Lord Jesus Christ.*

2 Peter 1:2-8 ESV

The grace of Almighty Most High GOD instantly puts the wicked ungodly into the family of GOD forever and that grace is based on the suffering and death of HIS Son Jesus Christ. I just heard preacher Bill Bailey of Happy Gospel Church say on a *SonLife Broadcasting Network* program that grace stands for, "**G**OD's **R**iches **A**t **C**hrist's **E**xpense," and that perfectly resonated with what I have been writing about since the late 1990's.

Getting to Heaven or entrance into the kingdom of GOD can only be accomplished by putting faith in the cross, shed blood, and resurrection of Jesus Christ the Son of the Living GOD. Our only contribution to enjoy the benefits of what Jesus did for all of us is to simply say, "Yes Jesus, yes!" And, of course, "thank you LORD for making me your child."

Grace is the name of the new covenant (New Testament) which was made between GOD The Father and Jesus HIS Son and sealed by the eternal blood of the sacrificed Lamb of GOD. HIS mercy, grace, and compassion are everlasting and far reaching forever according to the Word's use of the phrase "to a thousand generations" and they have been covenanted – assured – secured – guaranteed – made certain – obtained – pledged – acquired – and promised to be available whenever we need them. We are invited, maybe even commanded, to run boldly to the throne of grace when we need to be washed and scrubbed clean of the filth of this world and the residue behaviors of the sin nature which may flare up now and again. There is no excuse to remain spiritually dead, but it is your prerogative. The thing is there are no options for in-between. You either keep your death or you choose HIS life.

Do you not know that the unrighteous will not inherit the kingdom of God? Do not be deceived: neither the sexually immoral, nor idolaters, nor adulterers, nor men who practice homosexuality, nor thieves, nor the greedy, nor drunkards, nor revilers, nor swindlers will inherit the kingdom of God. And **such were some of you.** *But* **you were washed**, *you were* **sanctified**, *you were* **justified in the name** *of the Lord Jesus Christ and by the Spirit of* *our God.*   1 Corinthians 6:9-11 ESV

Wormholes of grace and mercy open when we confess our sins and brokenness. When we admit that we are sinners by birth, born spiritually dead, and willing participants of sin as evidence by our deeds and ask Almighty Most High GOD to forgive and cleanse us; we instantly become the children (Romans 4:4-5) of the One True Living GOD and joint heirs (Romans 8:17) of HIS resources alongside Jesus Christ the Son of GOD. The worse of human depravity can be cleansed by the blood of Jesus' cross and the reality of His resurrection. How is that for turning the lights back on?

You can open a wormhole of pardon, grace, mercy, and forgiveness right now if you desire. You can simply ask Jesus to show you that HE is real and that HE will forgive you. HE has already proven it by dying on your cross for you in your place (Romans 3:23-25) to pay your sin debt in full and satisfy the holy and righteous standards of the One True Living GOD.

You also can use the following scriptural prescription for getting washed, cleansed, sanctified, and justified (according to 1 Corinthians 6:9-11) and use it as a prayer as millions upon millions have done over the centuries. It is from the tenth chapter of the book of Romans which is an excellent starting place for cerebrally gifted and challenged minds. I say "challenged" because intellect can get in the way. Knowledge can get in the way of the simplicity of GOD's grace. Well, here's the scripture:

The word is near you, in your mouth and in your heart" (that is, the word of faith that we proclaim); **9** *because,* **if you confess with your mouth** *that Jesus is Lord* **and believe in your heart** *that God raised him from the dead, you will be saved.* **10** *For with the heart one believes and is justified, and with the mouth one confesses and is saved.* **11** *For the Scripture says, "Everyone who believes in him will not be put to shame."* **12** *For there is no distinction between Jew and Greek; for the same Lord is Lord of all, bestowing his riches on all who call on him.* **13** *For "everyone who calls on the name of the Lord will be saved."*   Romans 10:8-13 ESV

The book of Romans can be used as a sort of legal argument in the "courts of Heaven" to remind yourself and the kingdom of darkness that you belong to the family of GOD and no longer are you a child of damnation (Ephesians 2:3/2 Peter 3:7) because your biggest challenge in the beginning is convincing your mind that you are new. The residue of the sin nature will be used to convince you that you still are what you used to be before you asked your Heavenly Father to forgive you of your sinful deeds and to cleanse you of the sin nature.

Think of the book of Romans, and the entire Word of Almighty Most High GOD as a legal document and policy manual even which you can use to declare your rights and prove your innocence by the legal decrees of the Supreme Judge of the Ages. I know it sounds too good to be true, but it is true. And GOD put HIS life on the line to prove to you that it is true. And simple. As simple as looking and living according to Jesus' discourse with the top religious teacher in Israel at the time.

[Jesus said to Nicodemus,] *"As Moses lifted up the serpent in the wilderness, even so must the Son of man be lifted up that whosoever believes on him should not perish, but have eternal life.* John 3:14-15

Jesus was referring to an event which is recorded in Numbers chapter 21 in which the serpents which stalked the Jewish Hebrew Israelites during their wilderness journey were no longer held back by the protection of the LORD when the people rebelled against Moses. Everyone who complained and criticized Moses' leadership were bitten by deadly "fiery serpents." The LORD instructed Moses to fashion a bronze serpent and put it on top of a tall pole. Once the bronze serpent was erected, the victims of the fiery serpents only needed to look up at the bronze serpent to be healed of the poisonous venom which was coursing through their veins. Just look and live.

What agreement has the temple of God with idols? For we are the temple of the living God; as God said, "I will make my dwelling among them and walk among them, and I will be their God, and they shall be my people. 2 Corinthians 6:16 ESV

CHAPTER FOUR

THE HOUSE OF GOD?

And he was afraid and said, "How awesome is this place! This is none other than the house of God, and this is the gate of heaven!"

Genesis 28:17

Jacob used a rock for a pillow and as he slept, the angels of Almighty Most High GOD travelled up and down what some call Jacob's ladder. How can you mention a rock in a Bible story and not think about Jesus-the Rock of Ages? That rock which followed the children of Israel in the wilderness, we're told in 1 Corinthians 10:3-4, was Christ Himself?

Then Jacob awoke from his sleep and said, "Surely the Lord is in this place, and I did not know it." **17** *And he was afraid and said, "How awesome is this place! This is none other than the house of God, and this is the gate of heaven!"* **18** *Then Jacob rose early in the morning, and took the stone that he had put at his head, set it up as a pillar, and poured oil on top of it.* **19** *And he called the name of that place Bethel; but the name of that city had been Luz previously.*
Genesis 28:16-19

A possible wormhole opened above Jacob while he was sleeping and, in his dream, it was represented by a ladder. A direct connection between GOD and this man was at work. It would be some twenty years or so in Genesis 32 before we witness Yaakov/Jacob having another "wormhole moment" in which he "**wrestles with GOD"** according to Hosea 12:1-4. Even before leaving Laban's employ, Yaakov had another dream vision in which Almighty Most High GOD introduced Himself as **The** GOD of Bethel and instructing Jacob that it was time to go back home. Almighty Most High

GOD reminded Jacob of his dream vision from 20 years earlier with, "I AM The GOD of your Bethel experience." Isn't it great to be reminded that GOD is always there even when we don't sense or feel HIM?

Between Laban's home and Jacob's old home would be another wormhole moment. He would have to face Almighty Most High GOD. And HE would make Jacob face himself. Yaakov's name basically meant schemer, conniver, plotter, and even conspirator. His name represented the worse traits of the human temperament. In his encounter with GOD, Jacob had to speak his own name [in the original tongue] and all that it meant came pouring out of him. The *principle of confession* paints the picture of regurgitation. Confession has a cathartic effect. It is a cleansing of the soul and spirit. When we confess our wrongdoings and even hidden sins (iniquity), we purge ourselves of the residual filth which accompanies those imaginations and actions.

So, Jacob had to confess and in doing so, he had flashbacks of all the schemes he implemented over the years as well as the ones he only envisioned. In the presence of GOD, Yaakov had to face himself but once he vomited all of that stuff out by confession; Almighty Most High GOD could put something new on the inside. The "something new" which GOD put inside Jacob was memorialized by the wrestling match's souvenir. The limp. Before *The Cross* the most inner change one could get [my words] was a new mental image reminding them of an external experience. By that I mean that the resurrected human spirit core which Ephesians and Galatians proclaim could not happen until Yeshua completely atoned for the sins of humanity. After that anyone who put faith in the cross, blood, and name of Yeshua Jesus would have their dead human spirits resurrected "WITH CHRIST" according to Romans 6:4, Galatians 2:20, and Ephesians 2:6 to name a few.

Then Jacob was left alone; and a Man wrestled with him until the breaking of day. Now when He saw that He did not prevail against him, He touched the socket of his hip; and the socket of Jacob's hip was out of joint as He wrestled with him. Genesis 32:24-25

WHAT JACOB GOT WAS A LIMP. A limp which reminded him of his encounter with Almighty Most High GOD. The limp reminded Yaacov to respond to and introduce himself by his new name – ISRAEL. Wow, Jacob's wormhole experience changed the course of history, his people, and the world.

Eventually a wormhole of wrath, punishment, and indignation would open above the most important execution in human history. That wormhole moment would permanently solidify the eternal destinies of billions upon billions of people.

Jacob's wormhole experiences were 20 years apart. In the New Testament we will eventually see a connection between GOD and humanity which will never be broken or interrupted. And so, it was in his life. When Jacob had nothing at all he vowed to give The LORD a tithe of all that he receives. Is it possible that Jacob's Ladder was actually *Jacob's Wormhole*? It was so monumental that Jacob's memorializing his encounter with Almighty Most High GOD permanently changed the name of the city. The encounter also permanently changed Jacob. The "gateway of heaven" (Genesis 28:17) sounds like an event horizon, doesn't it? Maybe my past obsession with the sci-fi show Stargate influenced that last statement.

Interestingly, Jesus refers to Jacob's ladder when He meets His soon-to-be disciple, Nathanael, who also was known by the name Bartholomew. Jesus demonstrates His connection to GOD's Spirit when He proves to Nathanael that He knew what Nathanael was doing just before Andrew convinced Nathanael to come meet Him who Andrew believed was the Messiah.

*Nathanael answered and said to Him, "Rabbi, You are the Son of God! You are the King of Israel!" Jesus answered and said to him, "Because I said to you, 'I saw you under the fig tree,' do you believe? You will see greater things than these." And He said to him, "**Most*** *assuredly, I say to you, hereafter you shall see Heaven open, and the angels of God ascending and descending upon the Son of Man."* John 1:49-51

Ascending and descending angels? Is that the same language which Jacob used to describe his vision? Just as the angels of Heaven ascended and descended in Yaacov's dream vision, they did it constantly in Yeshua's/Jesus' earthly ministry. In the wilderness testing following His baptism, Jesus' use of the Word of GOD to overcome the adversary's temptation resulted in angels descending to serve Him.

Maybe in the same way an angel cooked for Elijah twice when he was fleeing the threats which Jezebel [Jeze-Baal] made regarding taking his life. The angels descended and ascended for Elijah to support his plight.

[Elijah] *lay down and slept under a broom tree. And behold, an angel touched him and said to him, "**Arise and eat**." And he looked, and behold, there was at his head a cake baked on hot stones and a jar of water. And he ate and drank and lay down again. And the angel of the LORD came again a second time and touched him and said, "**Arise and eat, for the journey is too great for you**." And he arose and ate and drank, and went in*

62

the strength of that food forty days and forty nights to Horeb,
the mount of God.   1 Kings 19:5-8 ESV

After the angel fed Elijah the second time, the prophet
was nourished and strengthened enough to last 40 days
and nights off of those meals. Wouldn't it be great to
have those recipes? The money we could save on our
food budget alone would be a blessing.   Anyway, after
40 days and nights of travel, Elijah ends us at Mount
Horeb which is also known as Mount Sinai. Yes, that
same Mount Horeb where Moses looked into the
glorious splendor of the GOD Of Creation and
"possibly" travelled through an Einstein-Rosen bridge
over 1300 years into the future and nearly 40 miles
north to Mount Tabor.

Then spoke Joshua to the LORD...and said, "Sun and moon stand still...and the sun and moon stood still...until the people had avenged themselves upon their enemies. So the sun stood still in the midst of heaven, and hasted not to go down about a whole day.   Joshua 10:12-13

CHAPTER FIVE

THE
LORD-OF-TIME
TECHNOLOGY

Adonai said to Moshe, "I will also do what you have asked me to do, because you have found favor in my sight, and I know you by name." But Moshe said, "I beg you to show me your glory!"

Exodus 33:17-19 CJB

There is evidence of supernatural time events (temporal singularities) throughout the Bible. The tenth chapter of Joshua shows an amazing occurrence. A man commands the sun and moon to stop moving and they obey. It is the same man who, in Joshua 1:8, YHWH instructed to meditate on HIS Word constantly so that he could realize its power. The tenth chapter of his story shows him doing just that.

The chosen people of Israel were in a battle and Joshua did not want to complete the day without a victory. The obstacle was the day was coming to an end and Joshua wanted more daylight so that they could keep the momentum which might be lost at nightfall. Joshua speaks to a star (the sun) and two planets (the moon and Earth), and time stopped as it regarded the courses of these celestial bodies. Interestingly, if you visit the Catholic Online site (catholic.org) you can find an article regarding the archaeologists who found an ancient Egyptian inscription which mentions Joshua's "wormhole" moment and dates the event on October 30, 1207, BCE. The Egyptians described the even as an eclipse. Fascinating.

The human soul and spirit are designed outside of time (Jeremiah 1:4-5) where Almighty Most High GOD exists. Imagine a gracious and compassionate being who is not and cannot be bound by time and space. Unless of course HE decided to live in a human body so that HE could teach us to be like HIM. Is that what happened? Is happening? Anyway, in the first chapter of Jeremiah's prophecy THE LORD tells the prophet that HE knew Jeremiah be HE formed him in his mother's belly. The spiritually dead say that life begins when the preborn takes their first breath. Hilary Clinton says that until then, "The preborn child has no constitutional rights." The spiritually alive say that life begins at the point of conception, and they are correct only in regard to natural life. The Word of Almighty Most High GOD says that life begins before conception. Connection with the Spirit of GOD begins before the egg is fertilized. The moment conception occurs, though, the countdown clock begins for every one of us. The countdown to death and the day of eternal destiny.

Also, don't the unalienable rights which the Declaration of Independence insists were endowed by the Creator and set forth in the Constitution of the United States of America actually come from Almighty Most High GOD (you know…Creator and All) and not from the constitution? Isn't the constitution simply the tool by which the founders implemented to communicate those unalienable rights? Spiritually dead political operatives with legal degrees are plagued by their sin nature as evidenced by their obvious juris imprudence. Again, the "rule of law" is spouted at when its conveniences them but their obvious hatred for the righteous law of GOD, laws of nature, laws of morality, and the like is evident by their duplicity.

One of my all-time favorite pop culture entertainment dramas is a space adventure known as ***Doctor Who***. Actually, it is quite a bit more than a mere space adventure. The Doctor character is a traveler who traverses outer space, time, and distance, at times, as easily as an urban jogger traverses a cross walk. For example, a bride-to-be accidentally stumbles into the path of the Doctor and her wedding jitters make it easy for her to agree to accompany the traveler on a mission which he assures her would be completed before she needs to show up to the ceremony...the next morning. The odd thing about this adventure is that it takes thousands of years to complete, and The Doctor still gets the bride to the church on time. Thousands of years of history experienced in less than twenty-four hours. Where do you think those concepts came from? Could it be the Bible? Who knows? Anyway…

The Doctor gets the bride to the church in time, and it is a great and sunny day for a white wedding. The Doctor's transport, known as the TARDIS (**T**ime **A**nd **R**elative **D**istance **I**n **S**pace), looks like the British police box. This time/space/time ship is the size of a phone booth on the outside but that is not the intriguing thing. Inside the TARDIS the mouth of each new passenger drops to find as many levels, rooms, and hallways as the Star Trek Enterprise. Each new passenger responds in the same way. "It's bigger on the inside!" Every new-creation believer is bigger on the inside because Jesus was given the Holy Spirit without measure (John 3:34) and Jesus resides inside each and every true disciple of His.

The master plumber and evangelist Smith Wigglesworth has been credited with quoting repeatedly a saying like this or something to the effect of, "I'm a thousand times bigger on the inside than I am on the outside." The miracles which resulted from him praying for a myriad of conditions including death tend to confirm just that. Still, Jesus guaranteed us that the least in the kingdom of grace was greater than all of the Old Testament prophets; so how much bigger on the inside do we need to be to see Heaven flow through us like it did Wigglesworth?

When GOD does a time miracle inside you, it can have the effect of the fullness of your glorified body (in the future) doing the work in the past (right now). I don't mean to complicate anything but couldn't GOD who exists out of time and sees it in two dimensions (my thoughts) answer a prayer in our past which HE knows we will pray in the future? There is no time in the Spirit. Only now. Several examples of the hopeless condition of death being reversed by a person with the Spirit of GOD on the outside of them like a cape or cloak are found in the Old Testament.

What we usually miss in reading, studying, and preaching these scriptural accounts is the time miracles involved. The people who died as a result of a disease or seeming brain aneurism (2 Kings 4) had to be healed of what killed them. That time miracle occurred in conjunction with the person being brought back to life. In several science fiction dramas, the theme of darkness seems to invade every plot. The undead operatives of Torchwood are terrified by that "something in the darkness" after they are brought back from death by some magical technology. The Doctor is being stalked by the coming darkness known as Bad Wolf. Time is better spent exploring what the Word of GOD has to say about gross darkness, outer darkness, and spiritual darkness. It's time well spent.

All of that being said, I am sure you must wonder why I'm going on in such a way about a science fiction show when I'm supposed to be dealing with wormholes from Heaven. If you are familiar with the sci-fi genres, then you have a frame of reference for the discussion this book is presenting. Well, I think that this Bible study will show that dimensional shifts, time travel, wormholes, and traversing the space-time continuum in inexplicable ways were all experienced by Enoch, Moses, Elijah, pagan king Ba'alshazzar, Deacon Phillip, John the brother of James, and Jesus Himself to name a few.

These recorded adventures began well over 5,500 years ago. There are indicators in scripture which could lead one to conclude that Isaiah, Jeremiah, and Ezekiel experienced such events as well. We will even see a few examples in the New Testament. And we will take a closer look at those examples in scripture passages which you are already familiar with to find the wormhole factor.

TRADITION SAYS that Moses was born about 1392 BC. Over 1,400 years before Jesus transformed on what some believe to be Mount Tabor in the presence of his top three disciples in Matthew 17, Mark 9, and Luke 9. Moses came down Mount Sinai/Horeb in Exodus 34 with the reflection of GOD's glory on his face and beard. What caused that imprint on Moses' visage? It is not often that I get to use the word visage so that explains my use of that word. Anyway Let's fast forward into Moses' recorded feature for a clue.

Six days later, Yeshua took Kefa, Ya'akov and his brother Yochanan and led them up a high mountain privately. As they watched, he began to change form — his face shone like the sun, and his clothing became as white as light. Then they looked and saw Moshe and Eliyahu speaking with him. Matthew 17:1-3 CJB

Peter (Kefa), James (Jacob/Ya'akov), and John (Yochanan) were Jesus' inner circle so to speak. Moses (Moshe) obviously represents the law and Elijah (Eliyahu) the prophets, but Jesus is the fulfilment of the two-fold display of GOD's grace under the Old Covenant as well as their fulfilment in the New

71

Covenant of grace. Anyway, where did Moses come from? How did he know where to rendezvous with the Lamb of GOD? Did Moses simply stand still in the cleft of the rock as the goodness of Almighty Most High GOD passed by? Did seeing the matchless magnificence of GOD's trailing glory open a wormhole or cause a dimensional shift in which Moses turns around because he is hearing Jesus speaking with His top three on Mount Tabor? Is Almighty Most High GOD so far above and ahead of humanity (Isaiah 55:8-9) that HIS backside or "past" is thousands of years into our future? What a might GOD we serve!

I don't know but I am really curious why pop culture is flooded with images and ideas of time travel and quantum leaps. Where do you think those concepts came from? Do we even want to ask if Elijah's arrival on Mount Transfiguration happened immediately after his apprentice Elisha saw him blast out of orbit on his fiery chariot ride in 2 Kings chapter two?

Kefa said to Yeshua, "It's good that we're here, Lord. I'll put up three shelters if you want — one for you, one for Moshe and one for Eliyahu." 5 While he was still speaking, a bright cloud enveloped them; and a voice from the cloud said, "This is my Son, whom I love, with whom I am well pleased. Listen to him!" 6 When the talmidim [students/disciples] *heard this, they were so frightened that they fell face down on the ground. 7 But Yeshua came*

and touched them. "Get up!" he said, "Don't be afraid."
8 *So they opened their eyes, looked up and saw only Yeshua by himself.* Matthew 17:4-8 CJB

Apostle Paul's personal physician and disciple of Christ, Dr. Luke, interviewed as many people as he could to piece together a chronicle of the life and ministry of Jesus Christ. The first chapter of the book of Acts indicate that Luke may have been commissioned (funded) by a possible nobleman to gather as many details as he could about the carpenter from Nazareth. The original title for the gospel of Luke and the book of Acts was known as *The Chronicles of… [Christianity].*

Dr. Luke cites in his account of the transfiguration event that Moses and Elijah visited Jesus on Mount Tabor to discuss the Savior's passion and death which He would accomplish at Jerusalem. Thank GOD that mission was accomplished. Still, where did Moses and Elijah come from? How did they get to Mount Tabor for the transfiguration at just the right time?

And the Lord said to Moses, I will do as you say: for you have grace in my eyes, and I have knowledge of you by your name. ***18*** *And Moses said, O Lord,* **LET ME SEE YOUR GLORY**. ***19*** *And he said, I will make all the light of my being* [goodness/glory] *come before you, and will make clear to you what I am; I will be kind to those to*

73

whom I will be kind, and have mercy on those on whom I will have mercy. **20** *But it is not possible for you to see my face, for* **no man may see me and still go on living.** **21** *And the Lord said, See, there is a place near me, and you may take your place on the rock:* **22** *And when my glory goes by, I will put you in a hole in the rock, covering you with my hand till I have gone past:* **23** *Then I will take away my hand, and you will see my back: but* **my face is not to be seen.**

Exodus 33:17-23 BBE

Some scholars speculate that Moses moved forward in time and space during this interaction with Almighty Most High GOD to the top of Mount Tabor where Jesus' transfiguration occurred. Is it possible that the glory which was reflected on Moses' face came out of Jesus' face. Is it possible that the glory light which came out of Jesus' inner core is what imprinted Moses' face? Moses requested to see GOD's glory (splendor/essence/power) but such a request to witness any King's splendor implies a face-to-face interaction is the actual request.

GOD told Moshe that the experience of seeing HIS face would prove fatal to Moses so HE would only allow Moses to see HIS "hinder parts" or backside. Some scholars say that the words indicating the backside or trailing wake of GOD's glory could have the implication of "***the future.***" I cannot show you

74

my face, but I can show you the future. Is that what really transpired? OUR GOD IS SO AMAZING and powerful that HIS backside or what's behind him is still thousands of years into the future for mere humans.

IS IT POSSIBLE that while Moses was in the rock's crevice viewing the trailing wake of GOD's glory train that a relativity bridge opened transporting him through time, space, and redemption history? Moses was shown an aspect of GOD which, as mind blowing as it was, was still less than what Moses requested.

Do you get that? Moses' request to see the face of Almighty Most High GOD was denied but the consolation prize was so out of this world and bent reality in such a way that professional relativity speculators (theoretical physicists) would be left with their jaws on the ground in amazement and relative awe. And we don't see Moses making that request again. I wonder why?

And the Word became flesh and dwelt among us, and we beheld His glory, the glory as of the only begotten of the Father, full of grace and truth. John 1:14 NKJV

God *who commanded light to shine out of darkness,* **has shone in our hearts** *to give the light of the knowledge of* **the glory of God in the face of Jesus Christ.** *But we have this treasure in earthen vessels, that the*

excellence of the power may be of God and not of us.
2 Corinthians 4:6-7 NKJV

The glorious splendor and power or Almighty Most High GOD are flowing in us out of the face of Jesus Christ according to this scripture. Is that a reference to the Mount Transfiguration phenomena? **Is it possible** that the eternal power of GOD's Word being deposited inside the spirit core of the new creation believer makes each and every one of us a potential supernatural phenomenon? A wormhole going somewhere to open? Just as Ezra was called a "ready scribe" (Ezra 7:6) who could write out the entire collection of the scrolls if they got damaged; are the new creation believers wormholes-at-the-ready?

Is every believer intended to be a walking wormhole? How else are we supposed to think about these types of scriptures which paint pictures of what the cross, blood, name, and resurrection of Christ has accomplished in us? Images and pictures of the unlimited power of Almighty Most High GOD resident in us now and throughout all eternity. Right now, every true believer and disciple of Christ is as righteous and justified as we will ever be and a partaker of Heavenly Father's divine nature.

Lauren's Daigle's song *How Can It Be?* asks the perfect question about how lavishly the GOD of creation has poured out HIS love and grace upon humanity. Sadly, though, not everyone will accept it. Too many even in the church do not believe and receive the love and grace which our LORD is anxious to lavish on us. The Word tells us in the letter to the church of Ephesus that GOD's amazing grace will still be unfolding thousands of years [my words] from now.

So that in the coming ages he might show the immeasurable riches of his grace in kindness toward us in Christ Jesus.
Ephesians 2:7 ESV

Now, if HIS trailing glory is thousands of years in the future for us; imagine how much more grace HE has for us just because of what Jesus did to secure us a place in the family. What more can there be? Well, Jesus said that HE is the door and the way. Think about it. The way to a destination. The door is not the most valuable part of a house or home, is it? What in the unseen realm could there be that is more than our blood shedding Door and grace securing Way?

*Grace and peace be multiplied to you in the knowledge of God and of Jesus our Lord, as His divine power has given to us all things that pertain to life and godliness, through the knowledge of Him who called us by glory and virtue, by which have been given to us exceedingly great and precious promises, that through these you may be **partakers of the divine nature**, having escaped the corruption that is in* *the world through lust.*   2 Peter 1:2-4 NKJV

*But also for this very reason, giving all diligence, **add** to your faith virtue, to virtue knowledge, **6** to knowledge self-control, to self-control perseverance, to perseverance godliness, **7** to godliness brotherly kindness, and to brotherly kindness love. **8** For if these things are yours and abound, you will be neither barren nor unfruitful in the knowledge of our Lord Jesus Christ.*   2 Peter 1:5-8 NKJV

Do you see that? To faith can be added virtue and to virtue can be added knowledge and so on. All of these spiritual capacities can be enhanced and augmented. Don't miss the fact that these spiritual attributes are already resident in every new creation believer. It is our responsibility as **co-laborers** with Almighty Most High GOD, in the sanctification process, to build up ourselves in every aspect. Praying "*in the Spirit*," according to the 20th verse of Jude's brief letter may be a really good step to add to the strategy.

*Beware lest anyone cheat you through philosophy and empty deceit, according to the tradition of men, according to the basic principles of the world, and not according to Christ. For **in Him dwells all the fullness of the Godhead bodily;***   Colossians 2:8-9

I listened to a few lessons in Andrew Wommack's

Bible study series titled **Christian Philosophy** in which Wommack describes the erroneous way Christians think and believe about spiritual concepts and clear-cut scripture. How we perceive or think about GOD, Jesus, Holy Spirit or any spiritual notions can have adverse impacts on our Christian walk and eternal destiny if our personal philosophy is not aligned properly with the Word of GOD. The first message which Jesus preached according to the 4th chapter of Matthew's gospel was, "Repent because the Kingdom of Heaven is being extended to you." Jesus came to the most religious people in the world who worshipped His Heavenly Father and told them to "change the way you think" which is what the word "repent" means.

*The mystery which has been hidden from ages and from generations, but now has been revealed to His saints. To them God willed to make known what are the riches of the glory of this mystery among the Gentiles: which is **Christ in you**, the hope of glory.*   Colossians 1:26-27

In The Son of Almighty Most High GOD resided the fullness or totality of the godhead and this same Son resides inside the resurrected spirit core of every born-again believer. Are we clay pots able to have fellowship with the members of the godhead (1 John 1:1-7) because the Cross, Blood, and Name of Jesus upgraded our resurrected spirit core to house the infinite Father, Son, and Holy Spirit? The Word of Almighty Most

High GOD affirms every new creation believer with an assurance and guarantee that the Greater One lives inside all of HIS children.

You are of God, little children, and have overcome them, because **He who is in you is greater than he who is in the world**.   1 John 4:4 NKJV

The **Spirit Of Truth**, *whom the world cannot receive, because it neither sees Him nor knows Him; but you know Him, for He dwells with you and* **will be in you**
         John 14:17 NKJV

Now when He was asked by the Pharisees when the kingdom of God would come, He answered them and said, "The kingdom of God does not come with observation; nor will they say, 'See here!' or 'See there!' For **indeed, the kingdom of God is within you."**
         Luke 17:20-21

When Jesus encounters the demoniac of Gadera (Matthew 8, Mark 5, Luke 8), we learn that the human spirit core has an astounding capacity. The "lead" unholy spirit inside the demoniac told THE LORD that his name was Legion obviously because of Roman occupation. A legion was a troop which represented about six thousand Romans soldiers. One Jesus cast out six thousand demons including the lead demon and Jesus said that the least in the kingdom of grace would do greater works than the greatest of all the Old Testament prophets. We definitely have some catching up to do.

And a highway shall be there, and it shall be called the Way of Holiness; the unclean shall not pass over it. It shall belong to those who walk on the way; even if they are fools, they shall not go astray.

No lion shall be there, nor shall any ravenous beast come up on it; they shall not be found there, but the redeemed shall walk there.

And the ransomed of the LORD shall return and come to Zion with singing; everlasting joy shall be upon their heads; they shall obtain gladness and joy, and sorrow and sighing shall flee away.

Isaiah 35:8-10 ESV

CHAPTER SIX

THE

~~HIGHWAY~~

WORMHOLE

TO HEAVEN

I tell you the truth, those who listen to my message and believe in God who sent me have eternal life. They will never be condemned for their sins, but they have already passed from death into life.

John 5:24 NLT

The theme scripture for this chapter is giving us some amazingly exciting news. A solid assurance of eternal security is found not in what we do but is found by putting faith in what Jesus did for us. That being said, this chapter will be brief because its point has already been made. But not too brief.

The first message which the most religious people in the world heard from a wilderness preacher named John Baptist was, "REPENT!" The Word tells us that after John was thrown in prison for preaching against King Herod, Jesus took up John's message and began preaching the exact same thing, "REPENT!"

Repent for the Kingdom of Heaven is being offered to you. The word translated "repent" means to *change the way you think until your thinking aligns with GOD's way of thinking*. You might ask, "How can we know what Almighty Most High GOD is thinking?" We can start by reading HIS Word the Bible. It is chock full of HIS thoughts. Someone actually asked that question is church once while the preacher was delivering a sermon. "How can we know what GOD is thinking," was posed to the audience. No one answered. Maybe that story is another in a plethora of urban myths. There is no way that could happen. Even if only half of the people in attendance had a Bible with them there would be at least one person who looked at their Bible and thought that maybe that was the answer. A book full of GOD's thoughts. Yeah, there is no way something like that scenario happened. But it really did. I believe the person who told the story.

So, if the most religious people in the world needed to change the way they were thinking about GOD, themselves, and others then something must have really been wrong. And if the chosen people of Almighty Most High GOD who had a blood contract with HIM had to get their mindset realigned with the Word of GOD; just maybe we all need to examine ourselves (2 Corinthians 13:5) to make sure that we are aligned with HIS Word. We are even admonished to test our "spiritual" mindset against the Word to assess whether or not we are standing in the faith of Jesus Christ and ensure that we have not fallen into the religion trap of dependence on our own efforts of dead religious work.

*For [**Jesus**] was crucified in weakness, but **lives by the power of God**. For we also are weak in him, but in dealing with you we will live with him by the power of God. Examine yourselves, to see whether you are in the faith. Test yourselves. Do you not realize that **Jesus Christ is in you**?—unless indeed you fail to meet the test!*

2 Corinthians 13:4-5

Do you see that? Jesus, God in the flesh, lives eternally by the power of Almighty Most High GOD after first

85

being crucified in weakness. A born-again human sits at the round table of the Trinity? Maybe? I do not know whether it is round or not. Maybe it is a triangle but that is not the point. The God-Man [100% divine and 100% human] represents redeemed humanity in Heaven after first being crucified in weakness. But He is not weak anymore. The Word of GOD states that that GodMan advocates for us before the Judge of the Ages but also intercedes for us there. If that was not enough and it probably could be, but we are also told that we sit with Him and in Him in Heavenly places far above the darkness which tries to stalk us still.

*YHWH's mighty power was performed in Christ, when HE raised Him from the dead, and set Him at HIS own right hand in the heavenly places, **far above all principality**, and **power**, and **might**, and **dominion**, and **every name** that is named, not only in this world, but also in that which is to come: And has put all things under His feet, and gave Him to be the head over all things to the church.* Ephesians 1:19-22

*But **God, being rich in mercy**, because of the great love with which he loved us, even when we were dead in our trespasses, made us alive together with Christ (by grace you have been saved) and **raised us up with him and seated us with him** in the heavenly places in Christ Jesus.* Ephesians 2:4-6 ESV

Is mindset which these scriptures offer the type of thinking going through Smith Wigglesworth's head when he pulled a corpse out of the hospital bed and screamed at it, "Live!" or was it some traditional sweet-by-and-bye unbelief which dead churches have been feeding their people with for centuries? Wait until you get there to enjoy all that GOD has for you is what the sweet-by-and-bye poison offers. If the object of our faith and our customary mindsets are not in line with the Word of GOD, then our faith is contaminated. Poisoned even. Contaminated faith kills the wormholes and clogs the pipelines.

So, how do we pass the test of our faith mindset? How do we make sure that we are actually in the faith? How do we ensure that we live a deliberate dependence on the power of Almighty Most High GOD the same way Jesus does? We need to possess the mindset where we have the realization of a constant need for the Savior. We need to be as dependent on the power of the cross, blood, and name of Jesus every day as we are the first moment, we kneel at the foot of *The* Cross to have our sins washed away and the sin nature excised from the disconnected spirit core. Jesus told us that without Him we can do nothing. Not one thing that matters eternally. When we awaken every day, we need to say, "Ready Jesus?" If we awaken each day with the correct seated-in-Heavenly-places-in-Christ with a healthy but humble GOD-inside-me mindset then the kingdom of darkness will have the same feeling it had that Sunday morning over two thousand years ago when our crucified Savior, whom they influenced the governments of the day to execute, started stirring after being in the grave since sundown on Wednesday (Passover Sabbath). As in the creation account in the first chapter of Genesis, the evening and the morning were the first day. Thursday sundown was the end of day one. Friday sundown was the end of day two. Saturday or sabbath sundown was the end of day three. Jesus' siesta was just about done and the rising of the sun on resurrection Sunday marks the biggest regret the kingdom of darkness has ever known and still laments.

We impart a secret of God's wisdom, a secret which God decreed before the worlds began for our glorification. None of **the rulers of darkness** *understood this, for if they had, they* **would not have crucified the Lord of glory***. But, as it is written, "What no eye has seen, nor ear heard, nor the heart of man imagined, what God has prepared for those who love him"*   1 Corinthians 2:7-9

That second chapter of 1 Corinthians goes on to says that we have received the Spirit which is from GOD. Okay, but why would that cause the kingdom of darkness to regret killing Jesus? Colossians goes further and farther into this mysterious secret which YHWH knew that if the powers, mights, dominions, principalities, rulers of the darkness, and the wicked spirits in high places (Ephesians 1:20-21) had known that they would not have influenced the Jewish religious leaders, Herod, and Pontius Pilate to execute Jewish Jesus.

I became a minister according to the assignment from God to make the word of God fully known to you, a secret hidden for ages and generations but now revealed to his saints whom God chose to make known how great among the **Gentiles** *are the riches of the splendor of* **this secret is Christ is in you**, *the hope of glory.*   Colossians 1:25-27

In the fifth book of the Bible, YHWH says that Israel is HIS peculiar treasure. When HE demands that Egypt let HIS people go, HE tells Pharaoh that Israel is HIS firstborn son. The secret which GOD kept from everyone is the fact that HE was not only going to be Emmanuel to HIS chosen covenant people but that HE would also be Emmanuel living inside the grace covenanted Gentiles. The ones which the rulers of darkness thought were in the bag and that they owned them. If the rulers of darkness knew that killing Jesus would result in:

1   An army of Jesuses growing exponentially
2   Jesus inhabiting their Gentile slaves
3   Jesus working through them as if it was Him
4   Gentiles preaching Jesus to Jewish Israelites
5   Jews and Gentiles made a new race in Jesus

then they would have left Him alone. Imagine if you will that one Jesus -The Jesus- defeated the kingdom of darkness and won salvation for any human who wants it and made a public spectacle of the army of darkness according to Colossians 2:15.

He disarmed the rulers and authorities and put them to open shame, by triumphing over them in him. Colossians 2:15

How can we live by the power of Almighty Most High GOD as Jesus does. How can we possibly come close to the stature of Jesus? He earned the most excellent name according to Hebrews 1:4, Hebrews 5:8-10, and Philippians 2::9-11. The thirteenth chapter of the second Corinthians letter reminds the hearer and reader that the GodMan resides inside all believers. The GodMan is the highway to Heaven. Jesus Himself is the door, the ladder, the wormhole, and the way to Heaven and kingdom of GOD.

When Jesus first met His soon to be disciple Nathanael, also known as Bartholomew, He assured this disciple that the angelic phenomenon of Jacob's ladder would be a constant occurrence in His life and ministry and that the phenomena would be visible to the right eyes.

Nathanael responded to Jesus, "Rabbi, you are the Son of God and you are the King of Israel." Jesus answered him,

Wormholes From Heaven

*"Because I told you that I know your character and what you were doing and saying under the fig tree just before you came to meet me, you believe? You will witness much greater things than these. Of a truth, I truly promise that from this point on, **you will see heaven open**, and the angels of God ascending and descending upon the Son of man.* John 1:49-51

The Word tells us that Almighty Most High GOD was inside Jesus Christ doing the work of reconciling the world to HIMSELF (2 Corinthians 5:19). Does that mean that Jesus was a walking wormhole having GOD HIMSELF flowing through Him? Was the Man part of Him completely available for the GOD part to flow through Him?

Was Jesus simply a conduit for the compassion and loving care of GOD? When Jesus was "moved with compassion" as the gospels of Matthew and Mark put it at least a few times; was that GOD saying to Jesus that HE wanted to heal all the sick people in the multitude (Matthew 14:14) that day? When Jesus healed the leper at the foot of Mount Transfiguration, was that really GOD HIMSELF saying, "It is my will; be clean of your leprosy," and does HE want to do the same with us?

Consider Deacon Stephen's "face of an angel" as it was described in the sixth chapter of Acts when he was on trial before the Sanhedrin. Actually, he was preaching to the Sanhedrin. When we think of angles, we somehow picture some sort of light associated with the manifestation. Could Stephen have been exhibiting the same glory light which was exuding from Jesus Himself on Mount Tabor when Moses and Elijah dropped by to discuss Jesus' upcoming suffering and death in Jerusalem? Stephen's ~~stairway~~ …wormhole...to Heaven flowed out his face.

But he, full of the Holy Spirit, gazed into heaven and saw the glory of God, and Jesus standing at the right hand of God. And he said, "Behold, I see the heavens opened, and the Son of Man standing at the right hand of God." But they cried out with a loud voice and stopped their ears and rushed together at him. Then they cast him out of the city and stoned him… Acts 7:55-58 ESV

Stephen was killed that day with Saul of Tarsus giving the okay to execute him. Stephen did such a great job of letting the GOD part flow through the Man part that Jesus gave Stephen a standing ovation. So, what was this angel light which came out of Stephen's face?

A wormhole of GOD's glory presence maybe? Stephen was so connected to the *Living Bread* which he feasted on through meditating on Jesus' words and the Old Testament scriptures that when the Word of GOD was flowing out of Stephen's mouth the glory light of GOD's Spirit presence was flowing out of his face. And for good reason. GOD was up to something by doing it that way. Remember Saul?

The Pharisee **Saul of Tarsus** was present at Stephen's stoning and holding the garments of those who stoned Stephen while Saul was agreeable to the execution. Saul saw that light which came out of Stephen's face. Saul may have been present during Jesus' mock trials as well. He may have heard Jesus' voice. I said that the light flowed out of Stephen's face for good reason because it was the same light which would envelope Saul of Tarsus in the ninth chapter of Acts as he was traveling to Damascus to find Jews who had converted to faith in the blood of Jesus. Saul was charged with bringing them back to Jerusalem for execution.

Saul was determined to stop the heresy of Jews following the revolutionary teachings of Yeshua Nazareth (Jesus of Nazareth) who said that He came to fulfil the Law but in Saul's eyes He had destroyed it. Saul's eyes were about the change once he was knocked off of his donkey and enveloped in that glory light while, possibly, a strangely familiar voice spoke to him in the Hebrew (Acts 26:14) tongue, "*Saul, Saul, why are you persecuting Me?*" Saul was determined to stop the heresy that day and Jesus was determined to stop the persecution of the church from Saul. Saul submitted to that voice and that light; both of which I am convinced were strongly familiar to him. His response to this familiarity was, "Who are You Lord?"

The rest is history. GOD flowed through Stephen the same way HE flowed through Jesus, and it helped to change the course of history, the course of religion, and the salvation of Gentiles. That experience which Saul had was so critical to the spread of the gospel that THE LORD made sure to do it in such a way that it arrested Saul's attention. The recounting of Saul's Damascus Road experience causes great pause to anyone who reads it to this day. GOD's use of Stephen's angel face light to get Saul's attention led to my salvation and yours. GOD be praised!

Even before Stephen's angel face wormhole moment, there was Peter's shadow incident. Maybe it can be called Peter's wormhole of healing and demon chasing power. Let's take a look.

And through the hands of the apostles many signs and wonders were done among the people. And they were all with one accord in Solomon's Porch. 13 Yet none of the rest dared join them, but the people esteemed them highly. 14 And believers were increasingly added to the Lord, multitudes of both men and women, 15 so that they brought the sick out into the streets and laid them on beds and couches, **that at least the shadow of Peter passing by** *might fall on some of them. 16 Also a multitude gathered from the surrounding cities to Jerusalem, bringing* **sick people** *and those who were tormented by unclean spirits, and they were all healed.*

Acts 5:12-16 NKJV

For a Sunday School teacher to be talking about wormholes in the common Bible accounts which we have heard taught and preached several times over the decades is quite odd. I understand. I do get that, but it might be even odder for that same teacher to assert that it was not Peter's shadow which healed anyone might come across as inconsistent. I mean, if GOD is doing marvelous things through wormholes, then why not shadows?

Well, it's just my opinion. In the same way that the woman with the issue of blood (mentioned in Matt. Mark, and Luke) did not find any magical healing power in the tassels and "wings" of Jesus' prayer shawl. She understood what the wings and the blue thread meant and spoke her faith based on what GOD's Word said about those elements. Her mixture of the Word, her faith, her imagination, and life risking corresponding actions opened her wormhole of healing. Why do I say that it was not the fringes of Jesus' prayer shawl ("hem of His garment") which had the power in it? Because there were thousands upon thousands of people in that crowd that day which Jesus' disciples said were touching Him. Only one person out of that crowd touched Him with their faith. Faith will always open a wormhole from Heaven.

Be careful though. Fear and doubt flowing out of our mouths will do the same. Those words will open a portal from the kingdom of darkness. The first chapter of Jeremiah tells us that GOD watches over HIS Word to perform it. The cosmic counterfeiter, the devil, watches over the words of doubt and unbelief which come out of our mouths to perform those. Several Bible teachers have said that fear is the flipside of faith and works the same way. Just as faith comes by hearing GOD's Word, fear comes by hearing evil tidings and bad news. The most critical moment we all have in those times is what we decide will be the next words out of our mouths. Here is a great example.

When the man of God saw her coming, he said to Gehazi his servant, "Look, there is the Shunammite [woman]. Run at once to meet her and say to her, 'Is all well with you? Is all well with your husband? **Is all well with the child?"' And she answered, "All is well."**

2 Kings 4:25-26 ESV

If you read the Shunamite woman's entire interaction with the prophet Elisha in the fourth chapter of 2 Kings, you will be blessed if you pay attention to what is going on. Here, the woman had rushed to see the prophet because her only child, a boy, had died maybe an hour or so earlier. From the description of the son's medical emergency, he may have had a brain aneurism or a heat stroke while working in the field with his father. The father had a servant take the boy to his mother where he died on her lap around noon.

Not only did the mother refuse to tell the prophet's assistant that the boy had died when the assistant inquired after him, but she refused to tell her husband or any servants of his death. She only asked the husband to get some servants to ready a chariot for her so she could visit the prophet. Before she left the house, she laid the lifeless body of her son on the bed in the rehab addition which she and her husband added on to the house for the prophet to live and pray.

No one was the wiser. Why? Because she saw inside her heart, mind, and imagination a lively boy who was going to give her some grandchildren. The next words out of her mouth were critical. "All is well," was her response when the prophet's apprentice asked about the son. And guess what? She was right. All was well by the end of the day. The prophet raised the boy from the dead totally because the mother was careful not to speak what the facts were but what the truth was going to do to change the facts to be.

Instead of speaking what was on the outside she spoke what was on the inside and -you guessed it- a wormhole of life restoring power opened in that room and she took her boy and [gave him something to eat]. I assumed she fed him because Jesus instructed Jairus and his wife to feed their twelve-year-old daughter after

Jesus raised her from the dead. As a matter of fact, when news came from home that the daughter was already dead, Jesus stopped Jairus from saying one word. When Jesus heard what the servants told Jairus, He immediately said, "Do not doubt, only believe!" Believe what?

And behold, one of the rulers of the synagogue came, Jairus by name. And when he saw Him, he fell at His feet and begged Him earnestly, saying, "My little daughter lies at the point of death. **Come and lay Your hands on her, that she may be healed, and she will live.***"*

Mark 5:22-23 NKJV

Jairus had already said, "Lay your hands on her… and she will live." The synagogue ruler already spoke his faith and Jesus interrupted him at the moment in which the next words out of his mouth would have nullified what was already said. Isn't that interesting?

If we have bad physical diets, we are told that we could be killing ourselves with our forks and spoons. Spiritually, we kill things including situations and people with the words which come out of our mouths. At a recent Bible study one of the participants recounted a story where someone was in the hospital in a failing condition. The person at the beside was the spouse if memory serves. Maybe the wife. If anyone came in that room speaking one word of even implied doubt and unbelief, the wife kicked them out of the room with, "Don't be speaking that in his presence!"

You might think that is a bit harsh and over the top, but the sick person got well and enjoyed many more years of wonderful health.   In fact, read Dean Braxton's book In Heaven! if you get a chance.   Dean had died but his wife prayed him back.   Dean is a believer and was one when he died about 15 years ago. He went to Heaven, but Heaven could not keep him because his wife would not let them. Why?   Because she was using that name.   Jesus!   Wormholes of blessings or wormholes of cursing.   Which ones are your words opening?

Anyway, Peter's wormhole came from what he learned from Jesus after the resurrection.   The same night of the arrest and mock trials of Jesus, Peter promised Jesus that he had more love for the LORD than the other disciples combined.   He even assured Jesus that if it was necessary that he die right alongside Him.   Jesus tells us in Matthew 20 and Mark 10 that the people who "die" beside Him was already arranged by Almighty Most High GOD.   Jesus assured Peter that not only would he not die beside Him but that he would actually deny even knowing Him and that Peter would do it three times at the break of day.

Imagine what went through Peter's mind when he heard that.   He was, no doubt, the toughest of the bunch. Herod wanted to kill Peter later because he knew that Peter was the strongest amongst the Apostles **after the resurrection**. Peter meant every word which he promised Jesus, but he did not realize until after "Pentecost had full come" (Acts 2) that he did not have what he needed to keep those promises to Christ. In the 21st chapter of John's gospel, the resurrected Jesus, after eating breakfast, asked Peter three times if he really had more love for Him than the others. One time for each denial? GOD knows.

Jesus asked Peter twice if he had the GOD kind of love (agapao) for Him and more than the others. Peter responded both times that he loved the Lord like a brother (phileo). The third time Jesus asked the question, He used the word "phileo" as Peter had been responding with. I am convinced that Peter learned from that interaction that it is not his love for the LORD which brings him peace with GOD but understanding, believing, receiving, and accepting the agapao love (unmerited-can't work for) which GOD has for him is all the grace he needs to make it into the kingdom of GOD and into Heaven if he wants to.

When former big mouth Peter understood that Jesus left Heaven just for Peter, it changed everything. When Peter realized that GOD confined HIMSELF to a human body just so that Peter could get to know HIM better, it changed everything. When Peter got it through his thick skull that Jesus did not have any favorites but agape loved them all the same, it changed everything.   As John wrote in 1 John 4:16, "*We have known and believed the love that God has for us. God is love, and he who abides in love abides in God, and God in him.*"

Do you see it?   Peter became a walking wormhole of GOD's agape love.   The compassion which "moved" through Jesus now moved through Peter.   He realized that he only needed to see it and believe it.   He walked around thinking about how much GOD loved him and it was contagious.

Peter spread a little of that on the crippled man at the gate called Beautiful outside the temple and the man jumped and leaped all the way into the temple to learn more about GOD's agape love.   Later on, Peter spread that agape love on a multitude of sick and demon tormented people and **they were all healed**! The agape love of GOD wants everyone to be saved from sin, everyone to be healed of sickness, everyone to be free of demons, everyone to be financially healed and whole, and HE wants to do it through you if you are willing to cooperate.   And HE wants to do it for you if you are willing to receive it.

The twelfth chapter of 1 Corinthians speaks of spiritual gifts extensively.   The Word is speaking of GOD flowing through HIS people to bless other people. "The gifts of the Spirit are given to every [new creation believer] for the common benefit of all," according to 1 Corinthians 12:7. That really sounds like GOD has recruited every one of HIS new residences (Holy of Holies) to flow through so that HIS will and kingdom will be manifested in this world as they are in the unseen realm. The wormhole through which we will be transported to Heaven is already inside of every blood wash, cross bought, child of Almighty Most High GOD. Wow, we each get to be GOD's co-worker. Can you imagine that?

So, we've discussed Peter, Stephen, and Paul in regard to the wormholes from Heaven flowing through them or for them. Some scholars have what is referred to as doctrinal objections to Peter's epistles because, some say, "there is too much dependence on Paul's writings" especially in regard to Paul's letter to the church at Ephesus. What the religious and scriptural intelligencia fail to allow for is that the same Spirit presence which was in Paul flowed through Peter. That same glory light which flowed out of Stephen's face also flowed through the pens of Peter and Paul. The "Gifts of the Spirit" subject of the twelfth chapter of 1 Corinthians uses the term "self-same Spirit in referring to the manner in which GOD distributes HIS gifts amongst HIS children.

The same Holy Spirit who spoke and wrote through Paul does the same through Peter. No allowance is given by the intelligencia just as it was with some of the Early Church Fathers (theologians) for the supernatural element. The GOD factor was not a consideration. For example, religious and scriptural scholars assert that Isaiah's prophecy was written by two different people. The tone, mood, and voice (as they say) of the first half of the book of Isaiah are different from those of the second half.

And I said: "Woe is me! For I am lost; for I am a man of unclean lips, and I dwell in the midst of a people of unclean lips; for my eyes have seen the King, the LORD of hosts!" Then one of the seraphim flew to me, having in his hand a burning coal that he had taken with tongs from the altar. And he touched my mouth and said: "Behold, this has touched your lips; your guilt is taken away, and your sin atoned for."   Isaiah 6:5-7 ESV

Well, Isaiah saw an immediate need for a change in his life the moment he got a mere glimpse of the glory of GOD and HIS magnificent splendor. The forgiveness and cleansing which Isaiah requested from GOD were instantaneous but the transformation (let's call it sanctification) was gradual. Isaiah's mind, soul and brain caught up with his spirit eventually. Isaiah's ministry and words had to change him as much as it needed to change others. The closer Isaiah got to GOD, the more the flow changed. It changed him as well as those around him.

Isaiah's prophecy offers ample evidence, or at the least indication, why the prophet Isaiah who wrote both halves of the book were the one person his mother birthed. Paul reveals to us in 2 Corinthians 3:18 that our wormhole to Heaven is found in Jesus' face. The closer we get to HIS face through feasting on HIS Word, the closer we get to glory divine and become more and more of what we eat.

Again, our natural bodies are made out of what we eat. Before birth they are made out of what our mothers ate. After birth, they are maintained out of what we eat. What do you think our spiritual bodies are made out of? Words are spiritual food according to several scriptural implications as well as invitations by Christ Himself. HE wants our spiritual diet to be exclusively the Word of Truth but **what if we feed on lies**?

What if we regurgitate those lies into the mouths (spiritual ears) of others like baby birds are fed by their mothers? If we feed on the Word of Truth, then our speech would reflect that, and our mouths will feed our spiritual bodies an excellent diet. If, on the other hand, we have a tendency towards a diet of lies, deception, and unfounded "facts" (network & cable news) then our spiritual bodies are diseased and extremely unhealthy. Remember that words are spiritual and mental food, and you are what you eat and think. If we do not eat the Word, then something else is going to eat us. Eat up!

Wormholes From Heaven

With the fruit of a man's mouth his stomach will be full; the produce of his lips will be his in full measure. Death and life are in the power of the tongue; and those to whom it is dear will have its fruit for their food. Proverbs 18:20-21 BBE

CHAPTER SEVEN

THERE'S A WORMHOLE IN YOUR MOUTH

*If you confess **with your mouth** that Jesus is Lord and believe in your heart that God raised him from the dead, you will be saved. For **it is by believing in your heart** that you are made right with God, **and** it is by **confessing with your mouth** that you are saved.*
Romans 10:9-10 NLT

Remember the Shunamite woman of the previous chapter? She was very careful not to say to anyone that her son had died although it was an absolute fact. Even when she was in the presence of GOD, represented by the prophet Elisha, she would not say that her boy had died. What was she waiting for? She was expecting TRUTH (John 1:17, 17:17) to change the facts. Whoa! That was good. Somebody write that down…uh never mind. Anyway…

I heard the most interesting statement at Bible study last night. One of the attendees told the story about a doctor who, when presented a difficult condition to treat, instructed a patient to speak to their body. Others in attendance paused attentively for more details. It appears that the doctor, in essence, told a patient to not agree with the symptoms but to speak the opposite. "I'm improving," was the stock answer which the doctor equipped the patient with for times when others inquired about the patient's health condition. As unusual as that prescription might sound, the weird part of the story is that the doctor's instructions worked. Can you imagine your doctor telling you to, "repeat these positive mantras and call me in the morning" if you presented with a difficult to treat condition? I really believe that we all have the ability to open up wormholes of creativity with our mouths.

The natural inclination is to think that such use of

words or positive confession is outlandish. Really? The Word of GOD says in Romans 10:9-10 that the use of confession is necessary to secure eternal life by the blood, cross, and resurrection of Jesus. That's right.

With everything which Heaven planned and Jesus did to pay for the sins of the world past, present, and future; none of that matters if we don't admit that we are sinners, transfer our trust to Jesus, and accept His sacrifice as payment in full for our sins and to replace our sin nature with His nature (2 Peter 1:4) made available by grace.

My son, give attention to my words; Incline your ear to my sayings. ***21*** *Do not let them depart from your eyes; Keep them in the midst of your heart;* ***22*** *For they are life to those who find them, And health to all their flesh.* ***23*** **Keep your heart with all diligence, For out of it spring the issues of life.** ***24*** *Put away from you a deceitful mouth, And put perverse lips far from you.* ***25*** *Let your eyes look straight ahead, And your eyelids look right before you.* ***26*** *Ponder the path of your feet, And let all your ways be established.* ***27*** *Do not turn to the right or the left; Remove your foot from evil.* Proverbs 4:20-27

Look at verses 21 and 22 above. There is something in the Word of Almighty Most High GOD that will be healthy for the flesh and life for everything else, I guess.

Still, it appears that what those "secrets" or revelations are may not be obvious because the Word says that those benefits are for "those who find them." Interesting. Out of the heart flows life? The issues of life? Does that mean that the words of our mouths help form and even correct the concerns and conditions of our lives? In the same way the doctor's instructions to his patient to speak the proper words to get better did. Even disturbing diagnoses? So, what if we filled the mind (soul) and heart (spirit core) with the Word of Almighty Most High GOD?

Could we train the brain and mind to lean towards painting the pictures of faith and supernatural possibilities? What if we develop a routine of scanning the Word of GOD regarding any challenge we need to address? Would the tongue be more likely to speak as much as possible what Father GOD's Word says rather than the facts? For example, the miracles in the book of Acts changed crippled bodies of people born lame, prison circumstances and freed the Apostles, and even the time for Deacon Phillip to travel 23-40 miles in a matter of seconds. Didn't Jesus say that words can change situations and even nature itself?

I tell you the truth, you can say to this mountain, 'May you be lifted up and thrown into the sea,' and it will happen.

But you must really believe it will happen and have no doubt in your heart. **24** *I tell you, you can pray for anything, and if you believe that you've received it, it will be yours.* **25** *But* **when you are praying, first forgive anyone you are holding a grudge against,** *so that your Father in heaven will forgive your sins, too."*

Mark 11:23-25 NLT

That sounds like a really good result. If we speak what the Word of GOD says regarding a situation, then we can expect the situation to change. Jesus did not instruct any of His disciples to ignore the mountain or say that it does not exist but rather to expect the act of speaking truth accurately based on the Word of GOD to alter the situation. Joshua did not ignore the approaching darkness in the tenth chapter of his book. He spoke to the sun, moon, and earth and everything stopped so that he could win the battle before the day ended. And it worked…or did not work until the battle was won.

The tongue can bring death or life; those who love to talk will reap the consequences. Proverbs 18:21 NLT

We will reap the consequences or benefits of the words which flow out of our mouths. It is totally up to us.

Jesus told them. "I tell you the truth, if you had faith even as small as a mustard seed, you [would] *say to this*

111

*mountain, 'Move from here to there,' and it would move.
Nothing would be impossible."*   Matthew 17:20 NLT

It sounds like Jesus is saying that faith speaks. "If you had faith then you would say…"   On the other hand, Jesus' brother James says in James 2:20-26 that the presence of faith would result in corresponding action or work that is in line with the words coming out of our mouths. The following testimony which I heard on the radio several years ago is a great example. It was a long time ago, so the details are to the best of my recollection.   A jogger stumbled on train tracks and broke his ankle while running alone at night.

He prayed over his ankle while confessing scripture and then something special happened when he confessed that his ankle was completely healed. The jogger says that "something" [the voice of GOD] asked him, "If your ankle was completely healed what would you be doing right now?"   He told the "something" that he would continue jogging so the "voice" challenged him to do just that.

The jogger stood up and tried to walk and fell down. He kept confessing that his ankle was healed and stood up again and tried to run and he heard the bones sounding like they were breaking even more. The jogger was experiencing excruciating pain as he slow trotted ten feet and then another ten.   If he had stopped, laid down, and screamed for help; no one would have blamed him. But he would not have experienced what he experienced on the next stride which he took. The instant that broken ankle hit the ground with all of the jogger's weight on it; the pain was completely gone, the ankle was completely healed, and the jogger continued his exercise session. Too often when church people discuss the "faith without works" scripture, they reference religious activity rather than action corresponding to the words of faith which are being spoken.

*Brood of vipers! How can you, being evil, speak good things? For out of the abundance of the heart the mouth speaks. A good man out of the good treasure of his heart brings forth good things, and an evil man out of the evil treasure brings forth evil things. But I say to you that for **every idle word** men may speak, they will give account of it in the day of judgment.   For **by your words you will be justified**, and **by your words you will be condemned**."* Matthew 12:34-37

Jesus promises us that we will give account on judgment day for every idle (empty, powerless, reckless) word which proceeds out of our mouths. We can be sure that the words which proceed out of GOD's mouth are never idle or empty. In fact, they are full of spiritual nutrition according to Jesus' contrast, in the fourth chapters of Matthew and Luke, between natural food and the healthy spiritual food of the Word of GOD.

It would prove to be a beneficial practice if we learned as much of the Bible as possible so that when the pressures, cares, and challenges of life squeeze us on every side; to let the only thing which proceeds out of our mouths be the Word of GOD and words which cause a river of chain breaking, mountain moving, demon chasing power of the Spirit to wash away anything in our lives which does not honor GOD.

…no human being can tame the tongue. It is a restless evil, full of deadly poison. With it we bless our Lord and Father, and with it we curse people who are made in the likeness of God. **From the same mouth come blessing and cursing**. *My brothers, these things ought not to be so. Does a spring pour forth from the same opening both fresh and salt water? Can a fig tree, my brothers, bear olives, or a grapevine produce figs? Neither can a salt pond yield fresh water.* James 3:8-12 ESV

Because we can open wormholes with our mouths, we need to be especially mindful not to recklessly open wormholes of cursing, death, and destruction by speaking about the bad things which are on display in the natural. Instead, we need to ensure that all of the words which we speak come out of our spiritual mind rather than our natural one.

For those who live according to the flesh set their minds on the things of the flesh, but those who live according to the Spirit, the things of the Spirit. For **to be carnally minded is death,** *but* **to be spiritually minded is life and peace.** *Because the carnal mind is enmity against God; for it is not subject to the law of God, nor indeed can be. So then, those who are in the flesh cannot please God.*   Romans 8:5-8 NKJV

The natural or religious mind is antagonistic and hostile towards GOD. What comes out of our mouths is based on the images in our hearts and minds so doesn't it stand to reason that we need to paint those pictures with a spiritual mind bolstered by the Word of Almighty Most High GOD? Otherwise, death will be flooding out of our mouths. Reckless destruction will follow suit. We must be careful to guard the words which come out of our mouths. If we use words which are contrary to the Word of GOD such as words of fear, doubt, doom, gloom, and expecting the worst that could possibly happen; then we will be giving the minions of Hell permission to act on those words.

Keep your heart with all vigilance, for from it flow the [issues] of life. Put away from you crooked speech, and put devious talk far from you.   Proverbs 4:23-24 ESV

If crooked speech and devious talk are the seeds sown out of our mouths, what kind of harvest do you think that will yield?

Death and life are in the power of the tongue, and those who love it will eat its fruits.   Proverbs 18:21 ESV
Jesus called the multitude, and said unto them, Hear, and understand: not that which goes into the mouth defiles a person; but that which comes out of the mouth defiles a person.   Matthew 15:10-11

Whether the subject is the kingdom of darkness or the Kingdom of Light; they are both voice activated. If we speak words which perfectly align with the Word of Almighty Most High GOD then we will open a portal for HIS Spirit, power, and grace to flow through. The Word says that GOD's angelic agents respond and hearken to "*the voice of HIS Word.*" Please don't miss this point.

The Word does not say in Psalm 103:20 that the angels listen and respond to GOD's voice which there is no question that they do. The lesson we need to learn from the specific wording of this scripture is that GOD's angels listen and respond to the voice or voicing of the Word of GOD.

If we voice GOD's Word regarding our particular and specific situations, then the angels of Almighty Most High GOD will make sure that the words which come out of our mouths will be accomplished. The LORD says in Jeremiah 1:12 that HE attentively watches over HIS Word to perform it. Any of HIS angels, I am sure, have the delegated duty to do the same. So, what have you got to say for yourself?

God was in Christ making peace between the world and himself, not putting their sins to their account, and having given to us the preaching of this news of peace.

So we are the representatives of Christ, as if God was making a request to you through us: we make our request to you, in the name of Christ, be at peace with God

2 Corinthians 5:19-20 BBE

CHAPTER EIGHT

YOU ARE

GOD'S

CO-WORKER

For we are God's fellow laborers…
1 Corinthians 3:9a

We are GOD's co-workers? What can we contribute which is of any value? GOD does all of the work. HE simply needs us to listen and obey. HE needs us to be willing vessels of HIS agape love first and foremost because HIS power flows as a result of HIS loving compassion and tender mercy for everyone. Even those who hate HIM so much they will kill HIS people, manipulate others to kill HIS chosen people, and do anything to kill HIS people. Everybody is going to die anyway so why not die well? In Christ, we can "be" well and through HIM we can die well. Oh well. Let's get to it.

When Jesus' disciples asked Him for instructions on how to pray, one example He gave them was imploring Heavenly Father to establish HIS will on Earth in the same manner it is eternally established in Heaven. On the night which Jesus was born into a human body, the angelic troop declared, "Peace on Earth and [GOD's] goodwill towards humanity." That statement represents Almighty Most High GOD's original intent for creation. Goodwill, grace, and fellowship. The second chapter of Luke's gospel tells of a 91-year-old widow, Anna the prophet, who seemingly remained at the temple since the death of her husband 84 years earlier. The Word says that Anna [Hannah-grace] served GOD in the temple day and night with prayers and fasting.

On the day which Jesus was being presented to the

priests for circumcision, Hannah may have entered the ceremony during the circumcision. When she saw Baby Jesus, she told those in attendance that redemption had come to the earth. Just before Hannah came across the circumcision ceremony, a prophet named Simeon ("hearken-hear") held Baby Jesus and spoke some exciting words about this baby being the salvation hope of the world and "a light to lighten the pagan world." Simeon also said some disturbingly cryptic words to Mary about the pain she would endure in connection with Jesus' mission.

What does any of this have to do with being co-workers of Almighty Most High GOD? Simeon, as he was holding the baby, revealed that he had been praying to see the Christ Child /GOD Man before his apparent impending death. Simeon may have been as advanced in age as Hannah was and possibly much older. It appears that they both had been praying for years that Heavenly Father would establish on Earth HIS perfect will which was already established in Heaven. Together, Simeon and Anna are telling us as they told the gathering that day to "hearken to grace."

You see, our understanding of the significance of Jesus' specific prayer training instructions regarding "*inviting*" GOD to do on the earth what HE has established in HIS heart in conjunction with Apostle Paul's statement of believers collaborating [co-laboring] with the Creator of Light and Life simply

121

blows the petty religious mind out of our head so that there is nothing left but endless possibilities. Even endless impossibilities [the perspective of the religious/carnal/natural mind]. What is impossible in the natural is possible through the Spirit presence of YHWH flowing through us the way HE intends.

Jesus looked at them and said, "With man it is impossible, but not with God. For all things are possible with God." Mark 10:27

I think the American Standard Version (ASV) says, of Luke 1:37, "No Word of GOD is void of power." Almighty Most High GOD made it all so why does HE need to be invited to do what HE clearly wants to do in the earth? In the beginning, HE gave dominion and authority over HIS creation to humanity. Genesis 1:26-30 and Psalm 8 make it crystal clear that GOD gave the rule to mankind over the earth. We also learn from the Word of the fall of mankind in Eden.

That story in conjunction with what the Word teaches us in the sixth chapter of Romans helps us to understand that when Adam obeyed the words of the serpent (dragon-devil Revelation 12:9, 20:2) instead of the words of Creator GOD; Adam, Eve, and their authoritative dominion fell under the control of the serpent. To make a long explanation short; the only way for Heavenly Father to do what HE wants to do

in the earth must come from invitation from either the
serpent or a human. Is there any guess why Jesus was
anxious to become a human according to Hebrews
10:5-7? Because YHWH gave dominion, rule, and
authority over creation to humanity, it would be a
violation of HIS delegated authority to do whatever
HE wants in the earth without permission (prayer/cry
for help) from a human.

Now, imagine if YHWH had a human body. What
could the serpent do about that? GOD could come
and go as HE pleases and do whatever HE wants
without violating human freewill. Because HE honors
freewill, unlike the serpent, there still remains the need
for an invitation. Sadly, though, some believers made
the mistake of thinking that they were co-laboring
(following even) with men rather than with GOD
through HIS Son Jesus Christ. In Corinth, for
example, some believers subscribed to Paul's teaching
to the exclusion of Apollos', and some to Peter's
teachings to the exclusion of Paul's. When Paul heard
about that, he "nipped it in the bud," as deputy Barney
Fife frequently suggested to the Mayberry sheriff when
a schism arose.

*Who then is Paul, and who is Apollos, but ministers
through whom you believed, as the Lord gave to each one?
6 I planted, Apollos watered, but God gave the increase.
7 So then neither he who plants is anything, nor he who*

waters, but God who gives the increase. **8** *Now he who plants and he who waters are one, and each one will receive his own reward according to his own labor.* **9** *For we are God's fellow workers; you are God's field, you are God's building.* **10** *According to the grace of God which was given to me, as a wise master builder I have laid the foundation, and another builds on it. But let each one take heed how he builds on it.* **11** *For no other foundation can anyone lay than that which is laid, which is Jesus Christ.* **12** *Now if anyone builds on this foundation with gold, silver, precious stones, wood, hay, straw,* **13** *each one's work will become clear; for the Day will declare it, because it will be revealed by fire; and the fire will test each one's work, of what sort it is.* **14** *If anyone's work which he has built on it endures, he will receive a reward.* **15** *If anyone's work is burned, he will suffer loss; but he himself will be saved, yet so as through fire.* **16** *Do you not know that you are the temple of God and that the Spirit of God dwells in you?* **17** *If anyone defiles the temple of God, God will destroy him. For the temple of God is holy, which temple you are.*

1 Corinthians 3:5-17

Who is Paul? Who is Apollos? Who is Peter? Who died for you? Who went to the cross for you? Who went to Hell in your place? WE ARE ALL GOD's fellow laborers. We are all in HIS service. Neither Paul, Apollos or Cephas (Peter) started this salvation and extended this grace. Yes, we are all GOD's co-workers in sharing HIS love and grace and by doing so building the Kingdom of GOD in the hearts of others. We also partner with the One, True, and Living GOD to expand HIS Kingdom inside ourselves. It is called the sanctification process. Yes, we must cooperate with GOD to have as much of HIS Spirit presence, glory, and power flow thoroughly through us as possible. The sanctification process can be summed up in John Baptist's statement when Jesus came on the scene for the first time.

He must increase, but I must decrease   John 3:30

The sanctification process is our outside catching up with our new creation resurrected spirit core inside. A major part of HIM increasing in our mindset and self esteem (now grace esteem) is allowing the image of HIS love for us to grow constantly in our imagination and faith.   Just as Peter simply rested in the fact that Almighty Most High GOD loves him so much that if he was the only one Jesus needed to come to Earth to save that GOD would have made that happen.

Like Peter, we must receive, believe, and accept the love which GOD proved, through the redemption plan, that HE has for us. In the sanctification process, our perfect insides catch up with our outsides which eerily seem just like the old us. I, you, we get to work with GOD to make each of us the very best us we were meant to be. Can you imagine that? You should. I do.

Jesus said to His disciples, "Because of your little faith, you could not cast out that demon but I truly say to you, if you have faith like a grain of mustard seed, you will say to this mountain, 'Move from here to there,' and it will move, and nothing will be impossible for you." Matthew 17:20

CHAPTER NINE

AMBASSADORS OF CHRIST OR WALKING WORMHOLES?

As a result of the apostles' work, sick people were brought out into the streets on beds and mats so that Peter's shadow might fall across some of them as he went by. Crowds came from the villages around Jerusalem, bringing their sick and those possessed by evil spirits, and they were all healed.

Acts 5:15-16 NLT

The answer to both questions is yes! Every believer is an ambassador of Christ our King and because He is the fullness of the GODhead bodily (Colossians 2:9), Heaven can open a wormhole inside any believer at any time to accomplish the wonderful works of GOD. (Acts 2:11)   That theme scripture for this chapter is taken from the fifth chapter of **The Acts Of** [The Holy Spirit Through] **The Apostles** more commonly referred to as the book of Acts. For some forty days following HIS resurrection, Jesus visited with HIS disciples on many occasions.

In the fifteenth chapter of the first letter to the Corinthians, Paul mentions that well over five hundred believers watched as my Lord ascended into Heaven.[2] Ten days before the church was birthed on Pentecost. That day on which Heaven opened and poured Holy Spirit on and into the 120 believers in attendance at the "upper room" prayer meeting.

Jesus instructed them to conduct this prayer meeting before they ventured out to spread the good news gospel. But why?   They could testify in a religious court of the resurrection.   They had hands on proof that resurrected Jesus was not just a spirit or ghost as evidenced by Him eating in front of them and inviting them to touch His crucifixion wounds.

Why did they need to wait for the Spirit presence of YHWH which was confined to the Holy of Holies core of the tabernacle during the wilderness sojourn and then the Holy of Holies core of the temple once it was built? Had they not been saved by the crucifixion and cleansed by the resurrection? They had but there was more that needed to be done.

When an unclean (unholy) spirit (demon) goes out of a man, he goes through dry places, seeking rest; and finding none, he says, 'I will return to my house from which I came.' And when he comes, he finds it swept and put in order. Then he goes and takes with him seven other spirits more wicked than himself, and they enter and dwell there; and the last state of that man is worse than the first. Luke 11:24-26

Swept, clean, and tidied up is not enough. A new resident needs to take custody of the property. On Pentecost, that resident was the Holy Spirit and once He arrive; there was no vacancy in which the kingdom of darkness could return. As eternally awesome as salvation grace through faith is, there is another work of grace which millions upon millions of believers have opted out of. The baptism in Holy Spirit with evidence of speaking in a Heavenly prayer language.

The men and women who actually walked, talked, lived with, and preached Jesus needed the second work of grace before they "**turned the world right side up**." (Acts 17:6) For some reason, the church world at large ignores this part of the birth of the body of Christ as if it is no big deal. For this cause, most of the church is in some form of rebellion. Ambassadors do not have the luxury of sharing their own opinion. The do not get the green light to, on-the-record, say what they think about any issue they are tasked with addressing. A legitimate ambassador only speaks the words which their King, president, or government tells the ambassador to.

I am sure that we have highly educated leadership in the body of Christ and the church world who understand this. Still, they are too eager to speak of the baptism in the Holy Spirit in a way which is contrary to the Word of GOD. Jesus said that it is important for ministry and eternal life (John 4:14) but His church leaders, by word and or deed, say that it is no big deal. Rebellion, rebellion, rebellion.

Remember how chapter seven of **Wormholes From Heaven** opened with the exciting assurance from Jesus that His true convert disciples have eternal life already? John 5:24 has some encouraging news for the believer. So, why don't we ignore all of the trash talk from Hell and its minions and simply believe what the Father, the Son, and the Holy Spirit have to say to and about us?

Hell's trash talk is just that – trash – and it is designed to keep our minds focused on the trash of our past in such a way that we consider that trash is the condition of our present. The first century church believers who were there at the birth of the body of Christ learned that they needed the baptism in the Holy Spirit to be able to stand even when it meant dying for their faith in Jesus' cross, blood, and resurrection.

These guys had seen, interacted with, and touched the resurrected Jesus. They knew that HE had been crucified and that He had been buried in a borrowed tomb on that Passover Wednesday. They also knew that the first people to proclaim or testify that the dead man walked out of that tomb were the very Roman soldiers which killed Him. Those soldiers were posted at the tomb to guard it. Isn't that amazing that GOD's "sense of humor" (if I may) was clearly displayed in having HIS Roman enemies be the first to tell HIS religious enemies of the resurrection?

Still, that group of well over five hundred believers which witnessed the ascension of the resurrected Christ after He visited with them for forty days still needed to wait for the Comforter to come (Luke 24:49/John 14:16-17/Acts 1:4) before venturing out with the message, "He is risen!"

That many people attesting to the fact that they visited with the risen LORD and that HE actually ate food to prove that HE was not some disembodied apparition could have gone a long way but that is not the way Almighty Most High GOD wanted the message spread.

The good news gospel had to convince the hearts (souls for the purpose of this discussion) and minds of men, women, and children of the three things which Jesus told HIS disciples to preach.[3]   The resurrection, repentance of sins, and forgiveness of sins is the good news gospel and it had to be preached by humans but humans could not do the convincing. Holy Spirit had to.

Jesus said in the sixteenth chapter of John's gospel that Holy Spirit would convince people of sin, righteousness, and judgment.   HE also informed HIS disciples that Holy Spirit would reside inside those who believed in Jesus.   So, His instruction to not preach the good news gospel before Holy Spirit had come must mean that it is not the believer or preacher who convinces the ungodly wicked sinner to convert but Holy Spirit inside the believer (there's that wormhole again) helping them say the exact thing[4] which will convince the heart and mind of the listener.

Holy Spirit opens a portal inside us which is referred to as "rivers of living water"[5] (John 4) through which power loaded words, prayers of faith, and the very healing presence of Almighty Most High GOD flow. The Samaritan "woman at the well" in the fourth chapter of John's gospel got a mere sip of Jesus' water and went back into town and preached to everyone that the very Christ -Messiah- prophetic Son of David – was at the well. "Oh, come see this man," she extoled.

The woman then left her water pot, went her way into the city, and said to the men, "Come, see a Man who told me all things that I ever did. Could this be the Christ?" Then they went out of the city and came to Him. John 4:28-30

The woman who was so ashamed of her reputation of being so broken and promiscuous that she came to the well at the hottest part of the day to draw water dropped her water pot and ran to town to bring others out to meet Jesus at Jacob's well. Only the women drew water at the well; so, this woman who had five possible divorces[6] in her past and was currently shacking up with a fifth man forgot all of that and ran back to town to tell everyone that she met the Best Man, and you need to come hear Him.

The living water from Jesus touched the Samaritan woman and her sip touched her city. She was not ashamed anymore. She only thought about His water. Jesus opened a portal in this woman which fixed her brokenness and washed her shame away. Jesus stopped at Samaria just for her.

And many of the Samaritans of that city believed in Him because of the word of the woman who testified, "He told me all that I ever did." So when the Samaritans had come to Him, they urged Him to stay with them; and He stayed there two days. And many more believed because of His own word. Then they said to the woman, "Now we believe, not because of what you said, for we ourselves have heard Him and we know that this is indeed the Christ, the Savior of the world."　　　　　　　John 4:39-42 NKJV

Of course, it was not Peter's shadow which healed anyone or cast out the demons. It was Peter abiding in the shadow of Almighty Most High GOD. Peter learned from being in that presence on a daily basis for about three and a half years. Peter's "limp" made him understand the unspeakable love and compassion which Almighty Most High GOD had for him.

When I speak of Peter's limp of course I am referring to him denying the Lord three times when it seemed that Jesus needed Peter the most, but Peter needed to deny Jesus before he could get to where Jesus wanted Peter. The point where the agape love of GOD would move through Peter just like it did through Jesus. Peter got so close to Jesus through meditating on that love that compassionate mercy flowed from Heaven through Peter and onto that multitude.

Peter learned that it was not about doing for Jesus to prove his love but about being loved by Jesus to extend GOD's love. Peter accomplished the greater works Jesus foretold (John 14:12) that all believers would accomplish simply by yielding to GOD's Spirit.

Show me your unfailing love in wonderful ways. By your mighty power you rescue those who seek refuge from their enemies.   Guard me as you would guard your own eyes. Hide me in the shadow of your wings.

Psalms 17:7-8 NLT

On the night of the Last Supper, Peter turned his love for The Lord into a competition when he promised Jesus that even if the eleven other disciples abandoned Him in His time of suffering and trial that Peter would not only have Jesus' back but would perish with The Lord.   Peter meant every word of it but did not have that wormhole moment, yet which would cause his permanent "limp." The permanent limp solidifies the faith and confidence and there is nothing that can assuage the individual once they get their "limp." Thomas' "limp" was his doubt. His incessant need for practical evidence.

That doubt was obliterated when the resurrected Jesus appeared before Thomas and invited him to touch His crucifixion wounds and give Him some food. Thomas did not know how to doubt anymore. "My Lord and my GOD," was Thomas' response to that encounter. He eventually took the gospel of the resurrection to India where he gave his life as a good soldier of *The* Cross. Whether it is Thomas' limp or Peter's limp we consider; it is important to realize that we all have one.

Even the unchurched know the story of Peter's failure. Even more than that, Peter actually denied ever knowing Yeshua Jesus to people who didn't matter. That's right. The people who were gathered around the fire to warm themselves who accused Peter of being Jesus' disciple were not people who had authority to do anything to Peter if he admitted to knowing Jesus. Peter did not have what he needed to stand. Not yet. None of the disciples which fled following Jesus' arrest had what they needed to stand. The Old Covenant did not end until Jesus died on the cross.

The New Covenant began when Jesus rose from the dead. Interestingly, when Yeshua Jesus died and ended the Old Covenant; the temple curtain which kept the presence and glory of Almighty Most High GOD away from the people [for their own safety] was "***torn from the top to the bottom***." A wormhole of celebration opened above the veil, and **I think** that Almighty Most High GOD HIMSELF tore that veil so His presence and glory could finally break free from this enclosure and now prepare to relocate to a new enclosure.

HE wanted to be mobile again as HE was in the wilderness sojourn. The presence moved whenever the people moved. The presence stayed wherever the people stayed. Now, HE can be and is mobile in the tabernacles with feet. Human bodies. Peter was one of those enclosures (tents-tabernacles) and when he got close to sick people they got healed. Deacon Phillip was one of those enclosures and when he preached in Samaria a sorcerer got converted to faith in Jesus' cross, blood, and name.

Deacon Philip preached Jesus, healed people, and cast out demons as a new enclosure of Almighty Most High GOD's presence and glory. Peter and Philip were walking wormholes of GOD's compassion and love **and so are you** if you have surrendered to Jesus as Savior, Lord, and Master.

The moment Peter denied the Lord, he realized that he did not have what it took to keep the vow he made to Yeshua Jesus to die alongside Him. When Peter heard that rooster crow the third time, he denied knowing the Lord; Peter knew then that Jesus knew it all along. The post resurrection Jesus gave Peter, along with the rest of us, a lesson on competitive love. Religion, false humility, and the residue of the sin nature deceive us into thinking that we can work enough to make up for what may be lacking in us. Before *The* Cross and Holy Spirit baptism Peter thought that proving his love for the Lord would make him a Lord's favorite. That was before *The* Cross.

He said to him the third time, "Simon, son of Jonah, do you love Me?" Peter was grieved because He said to him the third time, "Do you love Me?" And he said to Him,

"Lord, You know all things; You know that I love You."
Jesus said to him, "Feed My sheep.   John 21:17

After *The* Cross and baptism in the Holy Spirit, Peter understood that it was believing and accepting the love which Almighty Most High GOD had for him that made all the difference. Jesus asked Peter if he loved Him with and unconditional love. The Greek manuscripts of the scriptures use the word "agapao." The Greek alphabet has many words for the word "love." One of those words is philia from which we get the word Philadelphia. The city of "brotherly love." Jesus asked Peter if he loved Him with an agape or unconditional love. That was the top love. That is the kind of love which trumps all other loves (if I may) and is definitely the winner.

"Do you really have the most love Peter?"

Is that what Jesus was really asking him? Peter responded with, "I love you like a brother Jesus," to which Jesus asked a second and third time, "Do you have the most love for me Peter?" For years I've read this account but never connected the dots until I heard someone recently say that Jesus asked Peter once for each time Peter denied ever knowing the Lord. Jesus' covenant wounds of crucifixion were mocking Peter's former declaration of love. They also helped Peter rest in the reality of GOD's love for him, abandon religious love performance, and the result was miraculous. The compassionate agape love of Almighty Most High GOD flowed through Peter in such a way that multitudes of people were healed, and demons ran out of those tormented and possessed people.

Let your conduct be without covetousness; be content with such things as you have. For He Himself has said, "I will **never leave you nor forsake you."** *So we may boldly say: "The Lord is my helper; I will not fear. What can man do to me?"* Hebrews 13:5-6

How did Peter get from denying ever knowing The Lord to getting just close enough to sick people to make them well? Now, those same demons of shame and regret which would have taunted Peter for denying Jesus when He needed Peter the most simply ran out of people when Peter got close. Peter understood that believing, accepting, and receiving the agape love which Almighty Most High GOD has for him is what made him right with GOD. What about you?

No amount of penance, busy religious work or competitive religious love could be enough to make one right with Almighty Most High GOD. Understanding, receiving, and walking in HIS proven love for us will do that. Even when we are struggling with vices, sin, and bondage. No matter what, Jesus will not forsake His genuine disciples.

To forsake is to ignore you even while standing right beside you. To forsake is to turn the back to. To forsake is to leave behind. To forsake is to no longer value you. Even in our sin, vices, flaws, and failures; GOD will not ignore genuine disciples. If we had a genuine transformation at the foot of Jesus' cross as Romans 10:9-10 reads (just one example) then we are the legitimate children of YHWH. To forsake is to treat you like you don't belong. Almighty Most High GOD says that HE will never turn HIS back on HIS children. HE will never cast out those who belong to HIM. Faith in *The* Cross, the blood, and the resurrection of Jesus make us belong. Forever. They secure that promise and our blessed hope of seeing Him as He is. Because when we do, we will be just like Him.

I hope you are getting this. **The Lord knows** that we will fail and yet HE promises to not forsake us. **The Lord knows** that we will sin and yet HE promises to not forsake us. **The Lord knows** that we have the residue of that dead human inside us and yet HE promises to not forsake us. The oil of HIS Holy Spirit inside our resurrected spirit core covers the smell of death. The smell of that old sin nature. The smell of rebellion and animosity (Romans 8:7, Ephesians 2:15-16) towards HIS holy standards. The stench of the sin plantation is smothered by that alabaster aroma which kept Jesus focused mentally to keep His "natural" from getting in the way of His "spiritual." He wants to do the same with us.

*"If you love Me, keep My commandments. And **I will pray the Father, and He will give you another Helper**, that He may abide with you forever THE SPIRIT OF TRUTH, whom the world cannot receive, because it neither sees Him nor knows Him; but you know Him, for He dwells with you and WILL BE IN YOU. I will not leave you orphans; **I WILL COME TO YOU**.*   John 14:15-18

Wait a minute! Yeshua Jesus says that Almighty Most High GOD will send another Comforter known as the Spirit of Truth and then He says that it will be Him coming to live in us. Is Jesus the Holy Spirit? Is the Holy Spirit Jesus? Where is GOD the Father in all this? In you! Aren't you paying attention? Please keep in mind that it is the perceived equality of the Father, Son, and Holy Spirit in the Godhead which causes some to fall into the false "oneness" theology.

The oneness theology insists that Jesus is the name of the Father, Jesus is the name of the Son, and Jesus is the name of the Holy Spirit. Please don't make that mistake. The expression and ministry of each of the members of the Godhead is important. For good reason. Pray for wisdom and understanding in this area.

141

You have heard Me say to you, 'I am going away and coming back to you.' If you loved Me, you would rejoice because I said, I am going to the Father,' for My Father is greater than I. John 14:28 NKJV

I am the true vine, and My Father is the vinedresser. John 15:1 NKJV

How precious is your unfailing love, O God! All humanity finds shelter in the shadow of your wings. Psalms 36:7 NLT

Have mercy on me, O God, have mercy! I look to you for protection. I will hide beneath the shadow of your wings until the danger passes by. 2 I cry out to God Most High, to God who will fulfill his purpose for me. 3 He will send **help from heaven to rescue** *me, disgracing those who hound me. My God will send forth his unfailing love and faithfulness.* Psalms 57:1-3 NLT

Holy Spirit had the future King David prophesy about the Heavenly assistance coming down to rescue HIS people. I am so close to saying that Holy Spirit is a conduit (wormhole even) which allows whatever Heaven wants to send this way to flow through HIM and if we genuine disciples are cooperating with HIM; HE will flow to help us and HE will flow through us to rescue others. How do you get the tangible things of the unseen realm of Heaven to manifest or materialize in the realm of the natural? Imagination, faith, and prayer maybe?

What about word confessions which align with the revealed Word of truth? Holy Spirit mixes the recipe of the Word, faith, imagination, hope, and the expressed will of GOD to manifest in this world what GOD desires to happen. Just think about all of the rehearsals, prophecies, and confessions of the Old Testament. Eventually, the Word confessions and rehearsals became (materialized-manifested) flesh and dwelt in our midst. How can you use your mouth for anything other than blessing? Otherwise, you could be digging your grave with your tongue. Not joking!

Now all things are of God, who has reconciled us to Himself through Jesus Christ, and has given us the ministry of reconciliation, that is, that **GOD WAS IN CHRIST** *reconciling the world to Himself, not imputing their trespasses to them, and has committed to us the word of reconciliation.*   2 Corinthians 5:18-19

We could conclude just from the first fourteen verses of the first chapter to John's gospel that Jesus was an obvious walking wormhole of Word, Light, Grace, Truth, and Glory. In human form, Yeshua Jesus was a walking wormhole with the power, wisdom, love, and compassion of Almighty Most High GOD flowing through Him every minute of every day. Even on the night of His betrayal and mock trial and the day of His crucifixion. All of the goodness of Almighty Most High GOD came flowing through Jesus.

While He was being arrested in Gethsemane, He put a severed ear back on the head of one of the temple guards who was transporting Him to His death.

I KNOW SOME SAY that it was Jesus' humanity side which asked for **The Cup** (Revelation 14:10) of GOD's wrath and punishment on sin and the sin nature to be "taken away from Him" so that He would not have to experience the death of His body. I am not going to argue that this line of thinking is incorrect, but I will say that the death of Yeshua Jesus' flesh was of little consequence to Him. Something more important had His attention.

It was the death of His spirit which was the issue. It would "disconnect" from the Father (spiritual death) just like Adam and Eve experienced the moment Adam joined his wife in rebellious agreement. The instant he did, their glorious spirit cores disconnected from the *Source of Light* and life (John 1:9) and their glory lights of GOD's presence died out. In Matthew 19:8, Mark 10:6, Mark 13:19, and John 8:44 ;Yeshua speaks of specific knowledge which He had "from the beginning." He spoke as though He was there before it all began. Don't the first two verses of John's gospel say that He was?   That they were?

In the 14th through the 17th chapters of John's gospel account, Jesus prays for all of His disciples including us.   While doing so, He peels back a bit of His humanity to reveal that He was, is, and will always be a walking wormhole of Almighty Most High GOD's glory.

And this is eternal life, that they may know You, the only true God, and Jesus Christ whom You have sent. 4 I have glorified You on the earth. I have finished the work which You have given Me to do. 5 And now, O Father, glorify Me together with Yourself, with **THE GLORY WHICH I HAD WITH YOU BEFORE THE WORLD WAS.**   John 17:3-5

And now, every new creation born again believer is a walking wormhole of Christ who is a walking wormhole of GOD.   Just as GOD was in Christ reconciling sinners to HIMSELF, HE is now inside every believer continuing the work.   HE wants every person on this planet to know of HIS great love and care for them.   HE wants them all to know that they do not have to spend an eternity in Hell separated from HIM because Jesus already tasted their death for them and paid the price for repairing the breach and closing the gap between GOD and man.

HE will, however, honor your freewill if you decide to reject HIS offer.

The very presence and essence of GOD's nature was in human form in Jesus so that our "limp" would remind us that we must answer to a new name. Our very bodies are our "limp" because they remind us that we jars of dirt (2 Corinthians 4:7) carry the magnificent splendor of GOD's presence inside our spirit cores. We must introduce ourselves as the new creatures which faith in *The Cross* made us.

Our "limp" is our very bodies. These jars of clay we call flesh are made up of dust according to Psalm 103:14. The Word of GOD tells us in 1 Corinthians 4:7 that what we have received of GOD is nothing we can boast about because it was not the result of our efforts. Jars of dirt cannot produce the glory of GOD. It then tells us in 2 Corinthians 4:7 that the magnificent excellence of GOD's Spirit presence in housed in these bodies made out of earth so that all of the glory and praise go to GOD.

And when Jesus went out He saw a great multitude; and He was moved with compassion for them, and healed their sick.   Matthew 14:14

How God anointed Jesus of Nazareth with the Holy Spirit and with power, who went about doing good and healing all

who were oppressed by the devil, for God was with Him.
Acts 10:38

Peter finally realized that his competitive love meant nothing to Heaven. Only one love would qualify him to enter the Kingdom of GOD and Heaven. Pebble (Peter's name definition) love compared to a mountain of amazing grace? Come on Peter! Once Pebble stopped trying to be a boulder; a mountain of soul saving, disease killing, demon chasing compassion and love flowed out of the Rock Jesus and through him. And GOD is no respecter of persons. (Acts 10:34)

So, let it flow, let it flow, let it flow.

When Heaven does anything through us crackpots, all of the credit goes to HIM. That is what made Peter's "shadow ministry" so powerful. And liberating. Peter would never run off at the mouth again.

Can you imagine that?

And the glory which YOU have given to Me I have given to them, so that they may be one even as WE are one; I in them, and YOU in Me

John 17:22-23a

CHAPTER TEN

THIS *LITTLE* LIGHT OF MINE?

For it is the **God** *who commanded light to shine out of darkness, who* **has shone in our hearts** *to give the light of the knowledge of the glory of God in the face of Jesus Christ.*

2 Corinthians 4:6

Let me get straight to the point. Almighty Most High One True Living Creator GOD is shining in our hearts (resurrected spirit cores) according to the theme scripture for this chapter. The glory light of HIS Spirit presence is shining inside each one of us who has been washed in the precious blood of HIS dear Son Jesus Christ. Look at the premise scripture on the page opposite the title heading for this chapter. Jesus received the glory light from ABBA Father and deposited it into every new creation believer.

There is nothing little about this light which Almighty Most High GOD has put inside the resurrected spirit core of every new-creation believer. I have already read this chapter and the amazing scriptures which it is based on, and I completely stand by that statement. Let's start with this next scripture:

But we have this treasure in earthen vessels, that the excellence of the power may be of God and not of us.
2 Corinthians 4:7 NKJV

What treasure? The light, presence, and glory of Almighty Most High GOD maybe? As was said in the previous chapter; jars of dirt cannot produce the glory of GOD.

But we are bound to give thanks to God always for you, brethren beloved by the Lord, because God from the beginning   chose   you   for   salvation **through sanctification by the Spirit** *and belief in the truth, to which* **He called you** *by our gospel,* **for the obtaining of the glory** *of our Lord Jesus Christ*

2 Thessalonians 2:13-14 NKJV

The good news ("gospel") is that the redemption plan of Christ was not intended to be limited to merely the saving of our souls by washing away our sins and eradicating the sin nature. All of the things which GOD intended, and which Jesus accomplished were for the sole purpose of putting Jesus' Mount of Transfiguration light and glory inside screwed up us. What makes GOD's grace so amazing is that HE does this while we are still trying to shake the residue of being "dead in trespasses and sins" off of us.

At times we are convinced by religion, ignorance, and false humility that we still are what we used to be. Before our blood bath at the foot of Jesus' cross. The truth of the matter is that we have a constant need for a Savior. We must be of the mentality that although the Word of GOD has assured us that we have passed from death unto life; we must walk out this Christian pilgrimage hand in hand with Jesus.

He is the way to the Father as He said. We establish a mindset of a constant need for the Savior by accepting and receiving the fire baptism of and indwelling sanctification ministry of the Holy Spirit.

EVEN WITH OUR FLAWS and failures GOD shines the glory of Jesus' nature, spiritual DNA, and Spirit presence out of us screwed up messes. But why?

But we have this treasure in earthen vessels, that the excellence of the power may be of God and not of us.

2 Corinthians 4:7 NKJV

We cannot repeat this scripture enough. It is a constant reminder that the treasure which GOD will allow to flow through us is totally HIS work and none of ours. That is what makes it easy to examine ourselves in general and in regard to partaking of the Lord's table – communion. We simply need to remind ourselves that it is all HIS work and none of ours. That really comes in handy during communion. The rulers of darkness will attempt to make us focus on the sins, vices, flaws, and failures or our past to convince us that we are not worthy to partake of the communion.

*Examine yourselves as to whether you are in the faith. Test yourselves. Do you not know yourselves, that **Jesus Christ is in you**? — unless indeed you are disqualified*

2 Corinthians 13:5 NKJV

The fact of the matter is we became right with GOD

because of faith in a blood covenant made between the Father and Jesus. If we could not contribute the tiniest iota to being made righteous, why do we think we are worthy or unworthy to commune with the Lord at HIS table based on what we have done or not done? The proper way could go something like this:

"Father, thank you for saving me and for sacrificing your Son Jesus for my sins. Forgive me for any sins I am guilty of by doing what I should not or not doing what I should. I thank you that I am right with You because of what Jesus did for me on the cross. Jesus, I now eat Your flesh and drink Your blood to remind You that I believe, accept, and receive the love You have proven for me."

After such a prayer, simply eat and drink the communion. To improperly discern the Lord's body or eat and drink unworthily is to feel comfortable or assured that you qualify or are worthy to partake because you have been well behaved since the previous communion. The opposite is true as well. If you refused to partake of the communion because you have not behaved well then you do not properly discern the Lord's body. You cannot qualify yourself or disqualify yourself.

I am not saying that being indifferent about sinning, failing, and bondage is acceptable. Far from it! If you are comfortable with sinning and you do not "feel" remorse over, it then you probably are not a genuine disciple of Christ. If you truly want to live right but struggle in areas of your life and are "brokenhearted" if you sin, then the broken heartedness is a very good sign that your "want to" go saved when you knelt at the foot of Jesus' cross for forgiveness.

One reason faith is referred to as a fight by the Apostle Paul is that you have to fight with reason and logic. In the natural (carnal mind), it is understandable that one would feel like they are not a good Christian if they keep sinning after coming to Jesus; but the truth is once you come to Jesus, the tug of war begins. The tug between the spirit of the flesh and the Spirit of GOD.

Your flesh wants to do what it did before. Your new spirit core wants to do what the Spirit of GOD wants it to do. If you decide to side with the Spirit of GOD, then you will need to fill up on what GOD's Word says about you. It actually shows you what you are on the inside and once you learn that once you've been washed in the blood of Jesus, you already are what you are going to be when He returns. Once you establish that reality in your heart and mind then you are ready to fight. Until you do establish that reality you will feel like a boomerang going back and forth between light and darkness. GOD's treasure is in you now. Just do not boast about it as if it was something you did.

*"**I, Jesus**, have sent My angel to testify to you these things in the churches. **I am** the Root and the Offspring of David, **the Bright and Morning Star"***   Revelation 22:16

One of the things which John testified of Jesus is that He is in the midst of the candlesticks. The menorah. Jesus is the light inside His church. There is absolutely nothing we have done to produce that. No need or place for boasting or an arrogant posture.

What are you so puffed up about? What do you have that God hasn't given you? And if all you have is from God, **why act as though you** *are so great, and as though you* **have accomplished something on your own***?*   1 Corinthians 4:7 TLB

The Loving Father (John 3:35, 5:20, 16:27) has chosen to do HIS best work through screwed up people so that there is no iota of boasting on our part. On the inside, we are not screwed up, but our outside is quite a magnitude from that inner condition of righteousness. The good work which GOD had begun inside our spirit core (Philippians 1:6) is all HIS work and none of ours. The glory light which shines out of the darkness of the human psyche is proof positive that any accolades coming our way immediately go to HIM and not us.

When GOD was doing through Jesus the awe-inspiring wonders and miracles which are memorialize for more than two millennia now; could any of His disciples even imagine the same phenomena occurring through them?   It really should not be as difficult for us to accept as discussed in some church circles.   The power of Almighty Most High GOD flowing through the perfect Son of GOD is accepted without consternation.   Our Loving Father doing the same things through any other human is mind boggling and takes us aback.

The prospect of GOD working through others fills us with doubt.   The consideration of HIM working through us, in the beginning, fills us with dread until we come to the Biblical conclusion that GOD is so powerful and HIS grace is so amazing that HE can do anything HE says HE can do. Even work through me. Even work through you.   *THIS LITTLE LIGHT OF MINE* is actually HIS glory light shining as bright as HE wants it to through us.   That glory light with Jesus gave to us as HE said in John 17:22.

It is a testament and testimony of how powerful the blood of Jesus is and how strong GOD's grace is. The blood of Jesus washing our sin sick souls is what makes it possible for GOD to work in and through us to do HIS will. As we become increasingly comfortable with how GOD works through screwed up people so that all of the glory goes to HIM and none to us; then we will believe correctly. Believing correctly will result in us living correctly and behaving better.

"It is not about you being good enough but you believing in and depending on HIM who is GOD enough."

You see, GOD works from the inside out. HE isn't primarily concerned with behavior changing as HE is with a genuine transformation of the spirit core, mind, and emotions. As we begin to have increasing confidence in GOD's grace and goodness and decreasing focus on our worthlessness and "badness" then the glory light can shine brighter and brighter.

We must see GOD being GOD enough as the answer to our sometimes unspoken self-assessment about not being good enough. We must "see" GOD for who HE is and understand what HE desires for us.

The lamp of the body is the eye. Therefore, when your eye is good, your whole body also is full of light. But when your eye is bad, your body also is full of darkness. Therefore take heed that the light which is in you is not darkness. If then your whole body is full of light, having no part dark, the whole body will be full of light, as when the bright shining of a lamp gives you light. Luke 11:34-36 NKJV

What do you see when you look into the mirror of GOD's Word? Do you look into the mirrors of the law and the ministration of condemnation (2 Corinthians 3:9) and death and see everything which is wrong with you? Do you look at others through the legalism of *law-colored lenses* and fail to extend to them the same grace which the Loving Father has lavished on you? Do you look into the mirror of GOD's glory as described in the 3rd chapter of 2 Corinthians and see HIM changing you into the same glorious image of HIS Son?

In Luke 8:18, Jesus cautioned HIS listeners to be careful how they hear [the Word of GOD?]. Properly processing what we "see" in the Word of GOD as well as how we "hear" the Word of GOD is extremely important to setting the glory light of GOD totally free in and through us. At the same time, we should be cautious of everything we allow to enter our ear gate. What if the words we feed on are coming from a catalyst of anti-GOD and anti-Jesus sources? Will that nourish our souls and minds or poison them? You know, don't you?

That same light which "exploded" (my words) out of GOD when HE said, "ALEPH BET," (my assumption) has exploded inside every new-creation believer at the moment of salvation. When the wicked ungodly kneels at the cross of Christ to be forgiven of sins and wash in the blood of the Lamb of GOD, we instantaneously have the light of GOD's glory presence deposited inside our resurrected spirit core.

Although the church has sung the lyrics for decades, there is nothing little about the light which Almighty Most High GOD has invested in us. Has deposited in us.

Because it's HIM. HE's the light.

So, get after it and let HIM shine, let HIM shine, let HIM shine.

A Few Examples Of Possible Wormholes, Dimensional Shifts, and Temporal Singularities In The Bible

The Wormhole On The Original Book Cover

Cover

Jesus' Resurrection

John 11:25-26 Matthew 28   Mark 15:42-16:19   Luke 24

John 20 & 21 Romans 1:1-4   1 Corinthians 15:3-8

I trust that you have surmised that the image of viewing into a wormhole out of what looks like a tomb is totally connected to the resurrection of our LORD Jesus Christ.   The remaining wormhole examples which follow are in as close canonical order as possible. They follow the order of the events as spelled out in scripture.   An interesting activity which some of you might find intriguing is to get a chronological Bible to see Biblical history in the order in which it occurred.   Also, studying the resurrection from a scientific perspective might interest you.   For instance, I read one particle physicist surmise that the volume or air inside the tomb increased by 1 unit of atmosphere.   I am not going to pretend like I know enough about that to explain it but if 1 atmosphere is the normal and customary barometric pressure we experience on a daily basis,

imagine it doubling. More "existence" flowed into the tomb if I understand the scientist properly. Hebrews chapter one my help. It was the resurrection which was planned from before the foundation of the world that made all of the other wormholes possible. The Lamb of GOD was crucified "before the worlds began" is an important truth we glean from the scriptures. GOD planned it in HIS heart, so it was as good as done. The magnificent works of GOD (Acts 2) which occurred in the Old Testament were made possible by the crucifixion of the Son of GOD even before it happened at a fixed point in time and space. Because it first happened in the heart of OUR HEAVENLY FATHER. Whoa!

Spirit Of Man Breathed Into The Body   Genesis 1 & 2

The invisible parts, the spiritual bodies, of the first couple were created, given dominion, and had The Blessing spoken over them in Genesis 1:26-30 (day 6) but the first body, Adam's, was not made until either the 7th day [GOD's rest] or the 8th day of creation [the first day of the week]. Why would GOD [IF HE DID] create man's body on HIS day of rest? Because it was not work but worship or worthship. Man's value is exhibited in this possibility. Jesus said that humanity was not made to fit into the Sabbath (day of rest) but

161

that the Sabbath was made to benefit humanity. Anyhow, the spiritual body of Adam was not introduced into the physical body until the day (either 1 or 2 days later) GOD breathed that spirit into Adam's body. So, with an expulsion of "air," Adam's spiritual body traveled from the unseen realm into the tangible one.

The Decay Of Sin On Bodies Designed To Live Forever
Genesis 2, Romans 5, II Peter 3

The extremely long lifespans which are mentioned in the Bible have led the spiritually dead to conclude that the mention of such lengthy lives were allegorical or at the least, the length of a year was not quite the same as it is today. The fact of the matter is death did not enter creation until the fall of Adam. He was warned by GOD that **in the day** he sinned he would die. The cosmic day with the LORD is a thousand years according to Peter's revelation. Adam died physically 70 years shy of 1000 years of age meaning he did not cross into another "day."

Even the oldest man to die a physical death lived until he was 969 years old. Thirty-one years shy of 1000 years of age. Man was designed to live forever but sin opened a wormhole of death and decay. Science cannot explain why the human body "counts down" to death. The systems should operate indefinitely. The experts do not see what causes it to run down. That is because sin is invisible to the natural eye.

Abel's Blood Speaking   Genesis 4

In Genesis 4 and Hebrews 12, we are told that blood speaks. Praise GOD that the blood of Jesus speaks greater and better things than Abel's blood did.

Noah's Flood   Genesis 6, II Peter 3

There was not enough rainwater or underground water sources (Genesis 7:11, 8:2) which would have produced enough water which kept the planet flooded for forty days. A July 22, 2011, article on **space.com** website speaks of astronomers discovering "clouds of water vapor" surrounding black holes. Is it possible that the earth traveled through a wormhole through these clouds to augment the water production needed to flood the earth?

Maybe the galactic clouds traveled to earth via a cross-spatial or dimensional shift. I would also refer you to Michigan State University's site for an article on how galactic "rain" in black holes regulate star formation. The idea of hydrogen molecules making up these galactic clouds simply fell into my narrative quite conveniently.

Enoch's Cold Case?   Genesis 5, Hebrews 11, Jude 1

Enoch fathered Methuselah at age 65 and then fathered other children over the next 300 years. He had 365 years on this earth and then disappeared. It is told that "he was and then was not." Is that how the people who lived at that time would have explained what may have seemed like Enoch being "beamed up" like we witness in sci-fi movies? Was Enoch at a family gathering talking the same things he was talking for maybe 300 years? It is reported that he was constantly saying, "GOD is pleased with me." Did his faith confession open a wormhole for him to see the start of the Battle of Armageddon which is the second part of the Second Coming of Jesus Christ? He said of what he saw, "Behold the LORD returns with tens of thousands of thousands of HIS saints to execute judgment on the wicked." Did Enoch know too much to stay alive? No, he knew enough to have never died. Imagine what would happen if you said for 300 years, "GOD is pleased with me," so that others heard it. Would you be raptured just like Enoch? You might not have 300 years but why not start saying it now? I dare you.

Abram's/Abraham's Life
Genesis 12-18, Romans 4, Galatians 3

The man's entire GOD adventure was characterized by supernatural phenomena. Where to begin? Just help yourself.

Mrs. Lot Becoming A Pile Of Salt   Genesis 19, Luke 17

Did her disobedience result in a time progression in the mineral composition of her body?   How much time would it take a human body to petrify?   Would those minerals resemble salt?

Sarah's Retrograde Youth At Age 90 Qualifies Her For A Harem   Genesis 20

King Abimelech of Egypt took Sarah for his harem after Abraham told the king that she was his sister. He did so because the Egyptians apparently respected marriage and would not add a desirable woman to their harems if she were married. So, they solved their dilemma by killing the husband and imprisoning the widow in the harem. Less than a year earlier, GOD told Sarah that she would give birth at about 91 years of age. What did GOD's words do to Sarah's body to prepare it to carry and breastfeed a hungry baby boy?

Jacob's Ladder   Genesis 28

Jacob peered into the unseen realm and saw angels traveling up and down the ladder.   Did he get a glimpse through time and saw the face of his wife which would birth Joseph who would save the baby nation of Israel from destruction? He worked hard for fourteen years to see that face on a daily basis.

Nile River Water Turned To Blood   Exodus 7

In the first chapter of Exodus, the Hebrew Israelite male babies are thrown into the Nile on Pharaoh's orders because, "the people of Israel are too many and too mighty for us. Come, let us deal shrewdly with them, lest they multiply, and, if war breaks out, they join our enemies and fight against us and escape from the land." (Exodus 1:9-10)   The blood of millions of Hebrew Israelite baby boys over a span of at least forty years was added up and filled the Nile as well as every water source and vessel containing water at the time. Time Singularity?   Each of the ten plagues (might acts of judgment on the Egyptian cult religions) were designed to exhibit the supremacy of the GOD of the Jewish Hebrew Israelites in the minds of Abraham's descendants.

Water Flowing Through A Rock For ~ 4.5 Million People   Exodus 17

It really does not matter how many people there were. Getting water to flow out of a rock is definitely not natural.   The people were in need of water and on the verge of mutiny accusing of Moses of bringing them into the wilderness to kill them with thirst. GOD instructed Moshe to stand on a particular rock and strike it with his staff and HE would give them water. And HE did. Did the staff open a wormhole inside the rock or was there some rearranging of molecules and atoms in the rock to make water? GOD knows.

The Falling Of The Walls Of Jericho   Joshua 6

The walls reportedly were so wide that chariot races could be held on top of them. The shouts of praise either opened a wormhole through with the angels of Heaven came through to topple the walls of the Jericho fortress or Heaven multiplied the sounds coming out of the people's mouths to a resonance level which would cause stone and maybe bronze to crumble. Sound wave singularity? Was it a time event? Could the condition of the walls of Jericho 5K-10K years in the future materialize in that past and their present? GOD knows.

Joshua's Day The Earth Stood Still   Joshua 10:1-14

To make sure that this day of battle belonged to the Jewish Hebrew Israelites, Joshua spoke to the sun and the moon, and they stood still. We know that means that Earth stood still in its rotation and travel. The Catholic Online site, catholic.org has put up a video detailing the archeologist discovery of an inscription on Egyptian Pharaoh Merneptah's Stele which mentions an eclipse and how the day stopped until Israel defeated their enemies. The supposed date of the inscription is 10/30/1207 BCE.

Solomon's Guaranteed Mercy

II Samuel 7:15, Revelation 13:8

Because the Lamb of GOD was slain before the worlds were made, is it possible that GOD's guarantee to King David to never take away

HIS mercy from Solomon was based on the grace which the cross of Christ would bring in the future historically but had already happened essentially because it was GOD's best kept secret?

Bottomless: The Jug Of Oil & The Jar Of Flour
I Kings 17, James 5

The widow of Zarephath told the Prophet Elijah that she only had enough flour and oil for one last meal for she and her son and they were preparing to die afterwards. The Prophet challenged her to make a loaf of bread for him first and then for she and her son. He told her that if she did the flour and oil would not run out until the famine and drought were over. The Lord's brother James (Jacob) tells us in his letter to the church at large that the rain drought lasted three years and six months. How do you categorize that type of miracle? A wormhole of supply provision and increase? Whatever it was it definitely was not natural. It was definitely supernatural.

Fire From Heaven Falls On Elijah's Altar   I Kings 18

The Prophet Elijah confronted 450 false prophets as well as 400 pagan prophets on the top of Mount Carmel. Elijah presented a challenge to the fake prophets that they and he would build an altar. He would call on the GOD of the Jewish Hebrew Israelites and they would call on their deity called Ba'al. Actually, King Ahab's queen was named JezeBa'al (Jezebel) and these pagan and fake prophets worked for her. After the worshippers of the demon Ba'al prayed with no results; Elijah prayed and immediately Almighty Most High True And Living GOD answered with fire from the skies to burn the blood sacrifice. GOD's fire also lapped up the dust around Elijah's altar and the water in the trench which encircled it. This happened after Elijah said, "The GOD who answers by fire is the one we will serve. WOW!

Elijah's Departure From Earth On A Chariot Of Fire Pulled By Horses of Fire   II Kings 2

What Elisha described as a whirlwind could have been a wormhole. Picture a wormhole and think of the vocabulary which they had back then. You see it?

The Debt Destroying Jar Of Oil   II Kings 4

The wife of one of the members of Elisha's school of the prophets told the man of GOD that the debtors came to get her two children to place them in debtor's prison until the debts of the husband who recently died were satisfied. The Prophet asked her what she had of value and she said that she only had a jar of oil.

An Axe Head Floats To Pay A Debt   II Kings 6

The school of the prophets started building a facility for their ministry when a borrowed axe broke and the head fell into the Jordan

River. When the one who borrowed the axed let the boss Prophet know, Elisha threw a cut branch into the river near the place where the axe head fell and it floated. Now that it could be returned to the person from whom it was borrowed, the individual who borrowed it would not be indebted and nor would Elisha's school. A floating axe head? What kind of physics is that?

Elisha Completes His Ministry After Death   II Kings 13

King Joash had Elisha buried in an underground cave. As they were burying another man years (maybe) later; he was quickly thrown into Elisha's burial cave when a usual band of Moabite raiders was spotted coming to pillage and plunder as they did every spring. The man's body touched Elisha's bones and immediately revived and stood to his feet. Hopefully, they pulled him out in time to fight the Moabites. When Elisha saw his mentor Elijah transported in a whirlwind, the mantle (cloak/anointing) fell into Elisha's hands because he requested a double dose of what Elijah had. Elijah performed 8 major miracles in his life so Elisha, with a double portion of that anointing, should perform 16. At his death, Elisha had performed 15. Number 16 came when the dead man touched Elisha's bones (takes time to get have only bones left) and was raised from the dead.

The Earth Reverses 20 Minutes Of Day   II Kings 20

When King Hezekiah was informed by the Prophet Isaiah that GOD had answered the king's prayer for mercy and would extend his life by 15 years; Hezekiah asked the Prophet what sign he would get to know that the answer is correct. Isaiah asked the king if he wanted the sundial to advance ten degrees or go back ten degrees. In Hezekiah's estimation the easy thing would be to have the shadow on the sundial move forward by ten degrees (~20 minutes) so he requested the hard thing. The LORD granted the request through the Prophet and the "sun went back ten steps." And it did. GOD makes this stuff look easy.

Three Hebrews Thrown Into A Furnace   Daniel 3

Hananiah, Mishael, and Azariah told the king that even if GOD did not deliver them from the furnace which was heated up seven times hotter than ever that they would still not bow to his statue. The men were bound hands and feet and thrown into the furnace but when the king looked into the furnace he announced that the men were not bound, that they were walking around in the fire, and that a fourth man had joined them in the fire. The king said that the fourth man looked like "the Son of GOD." When the three Jewish Hebrew Israelites came out of the fire they did not smell like smoke or fire. Dimensional shift? Causality bubble of protection?

Babylonian King Ba'alshazzar's Disembodied Hand
Writing On The Wall   Daniel 5

The Babylonians had taken the Jews into captivity and the elements of the temple worship as well. The king and his royal circle were partying while drinking from the chalices which belonged in the Jerusalem temple when the hand appeared and began writing on the wall across from the menorah. Dimensional shift?

Isaiah Sees A Virgin Giving Birth   Isaiah 7

Bishop Fulton Sheen spoke of this on his *Life Is Worth Living* show several years ago showing that 700 years (if memory serves) to the day, Mary the Mother of Jesus Christ the Son of GOD gave birth to the Messiah. Did Isaiah only speak what he heard or did he see something indescribable?

Ezekiel's Wheel In The Middle Of A Wheel   Ezekiel 1

Is it possible that Ezekiel got a glimpse into the future even as far as the space age? Could he have been looking at a communication satellite? Read his account and think about it. Look at different images of NASA satellites and decide for yourself.

The Incarnation Of Holy Spirit Inside Believers   Acts 2

The Spirit of Truth, as Jesus called HIM in John 14,15, & 16 left the Holy Of Holies in the core of the Jerusalem temple and relocated to the spirit core of every new creation believer. When Jesus walked the earth, GOD was in one body. Since the Comforter has come to live inside Christians, HE has gradually grown in human temple residence. 120 believers in the upper room quickly had 3000 more added on after Peter preached in the street that day. Shortly afterwards another 5,000 souls accepted the cross of Christ for salvation and the mobile temple count has been growing exponentially. Hundreds of millions of temples on feet host the Holy Spirit and maybe well beyond one billion.

The Man Crippled From Birth Healed   Acts 3

The lame man who sat at one of the temple gates to ask money of those who were going into the temple was over forty years of age according to Acts 4:22. For someone who has never walked before to be yanked up by Peter and John was quite a risk. When the man's feet hit the ground his ankle bones and legs were completely healed including the tiny stabilizer muscles. That is an amazing thing. After the healing miracle the man enter the temple along with the Peter and

John "walking, and leaping and praising GOD," according to the record. A creative miracle maybe?

Peter's Shadow Healing The Sick And Casting Out Demons   Acts 5

Wow! Peter got so close to understanding Jesus' love for him that he abandoned the competitive love mindset and just let the agapao love of GOD flow through him onto others.

Wormhole Prison Break?   Acts 5

After the healing of the lame beggar at the temple gate named Beautiful Gate and multitudes of sick and demon possessed people got healed the religious council threw the Apostles in jail. The next morning the people who were sent to retrieve them from the prison cell returned to the religious leaders to tell them, "We found the prison securely locked and the guards standing at the doors, but when we opened them we found no one inside." (Acts 5:23) What? Yes! When you get a chance, please read verses 19-21 to see evidence of a wormhole moment. How did they get out without any guards or other prisoners seeing it happen? Did they move so fast that they passed through the walls and the people looked frozen-in-time to them? Was it a dimensional shift? Whatever it was, it was a circumventing of the laws of time and physics.

Missionary Phillip's Wormhole Travel   Acts 8

Phillip was doing some amazing work for the kingdom of Almighty Most High GOD when he was instructed by the Spirit to take a specific road which leads from Jerusalem to Gaza. After preaching to a Eunich from Ethiopian Queen Candace's qoverment and baptizing him, Phillip was instantly transported from the baptismal site near Gaza to a place called Azotus (current day Ashdod) which is twenty three miles away. Temporal singularity? Wormhole travel? Bilocation? Whatever it was it was definitely not natural.

The Conversion Saul of Tarsus (Paul)   Acts 9

The same light which Saul saw flow out of Deacon Stephen's face in Acts 6 and 7 enveloped Pharisee Saul as he was headed to Damascus to find Christians to bring back to Jerusalem for execution. The experience speaks for itself. Anyone can see the supernatural in the details of that experience. Look and see.

Dimensional Shift Prison Break?   Acts 12

After King Herod, the descendant of Esau, saw that killing Apostle James (Apostle John's brother) pleased the sporting amusement of the depraved spiritually dead people, he decided that he would have Peter killed as well. Peter was being held in prison and shackled between two soldiers and guarded by four squadrons of soldiers. I guess they heard of the wormhole prison break and was making sure that would not happen again. As you read the account you will find that Peter thought he was dreaming because he and the angel which led him out walked past both guard stations. The metal gate leading out of the city opened as if it had a modern day proximity sensor on it. After they got out of the city, the angel left Peter.

Paul Gets Killed, Gets Up, And Preaches The Next Day
Acts 14

GOD promised Paul something on Damascus Road that had not happened yet. Wow! Enough said?

Another Prison Break...But They Stayed   Acts 16

While in Philippi, Paul cast the demon out of a fortune teller who was also a slave girl which the leaders of the town exploited for financial gain for sure and political leverage as well no doubt. Paul and Silas are thrown into jail following the exorcism. Around midnight, the evangelistic team worshipped the LORD so much so that all of the prisoners heard them. The praise may have opened a wormhole which resulted in the subsequent earthquake which cause the cell doors to open. The prison keeper awoke at the commotion and was about to kill himself because he assumed the prisoners had all escaped. A Roman soldier or prison keeper who is determined to be derelict of duty was executed in front of their family. This man decided to save his family the optics and simply kill himself. Paul shouted, "Do yourself no harm because we are all here." The man, no doubt, heard the worship as he was going to sleep because when he realized that the missionary team not only did not escape but had such an impact on the other prisoners that no one did; he wanted that type of freedom. They showed that they were freer behind bars than the prison keep was on the other side of the bars.

Fatal Bible Study Accident And Raising From The Dead
Acts 20

Paul's lesson went a little long. Around midnight, a young man sitting in a window into a deep sleep and fell out of the window which was three stories up was dead when he was taken off the ground. Paul came outside and raised the young man from the dead. Maybe Paul's teaching covered the power of the resurrection that night and raising Eutychus from the dead was his opportunity to exhibit it. Whatever it was. It definitely was not natural. It was actually naturally supernatural. That is what the mindset of every believer should be. We should see ourselves as the walking wormholes we are. We should not be surprised when GOD flows through us to do amazing things in our lives as well as others.

Of course there are many more examples in the scriptures. I just wanted to make the point that the wonderful works of GOD (Acts 2:11) have and will circumvent the laws of physics and natural law if necessary as long as someone is courageous enough to believe what HE promises.

CLOSING THOUGHTS

Worthship Opens A Wormhole

Worship is happening all the time in Heaven.   When we get into a worship posture or mindset in our private times, we join with the family in Heaven to offer praise and adoration Almighty Most High GOD. When we assemble in our sanctuaries for corporate worship, we consort with worship which is already in progress in Heaven.   Praise and worship in the earth is the mouth of the wormhole which pulls on -makes demands of- the resources of Heaven.

But if you stay in me and obey my commands, you may ask any request you like, and it will be granted!   John 15:7 TLB

And I will do whatever you ask for in my name, so that the Father's glory will be shown through the Son. If you ask me for anything in my name, I will do it.   John 14:13-14 TEV

Extra! Extra! Read All About It
Deliberate Rebellion Opens A Wormhole In Wilderness

In Numbers chapter sixteen, a wormhole opened in the earth and caused at least 250 people to be transported "straight to the pit" (sheol-hell).   The phenomenon also swallowed the tents and possessions of those who rebellion against Moses and Aaron the High Priest. Take a look if you can.

If You See Something,
Say Something!

No matter how difficult the situations are which we endure, once we see the good, great, and precious promises in the Word of GOD regardless of how "real" the diagnosis and other physical evidence appears; then we must say what HIS Word says so we can receive what comes out of our mouths.

If we find ourselves saying what the obvious symptoms and conditions are then we are simply advertising for the darkness and devil.

The fact of the matter is you can agree with the facts, or you can agree with the truth and Jesus said in the seventeenth chapter of John's gospel that the Word of Almighty Most High GOD, "IS TRUTH."   When you see something in GOD's Word which will change your situation, repeat it. Enough said?

Tesseracts, Wormholes, Quantum Physics—and the Eucharist!

BY: Steven Greydanus (https://avemariaradio.net)

This article I found quite interesting. I will not include any of it here to avoid including too much of it and crossing some literary line or too little and causing more serious problems by sharing disjointed paragraphs and statements which should not, in my opinion, stand-alone without extensive explanation.

Please visit the site for the complete article.

The article makes the assertion that the ingesting the eucharist opens a wormhole of eternal life. Proceed with caution if you read it.

My Exception With The Catholic Assertion That Eternal Life Is Deposited With The Eucharist.

According to the Word of Almighty Most High GOD, eternal life is deposited into the resurrected spirit core of every new creation believer at the moment of salvation. It is not the act of ingesting the Eucharist which deposits eternal life into the spirit core of the believer because natural food cannot nourish deeply enough to benefit the unseen realm in which the spirit core resides. No, it is the unseen blood of JESUS which washes the sin-stained soul and dead human spirit core white as snow that makes all the difference.

IF INGESTING THE EUCHARIST resulted in eternal life then why wasn't the manna in the Ark of the Covenant simply multiplied and handed out? No, it is faith in the blood of JESUS' substitutionary sacrifice which

*But **Christ came as High Priest** of the good things to come, with the greater and more perfect tabernacle not made with hands, that is, not of this creation. Not with the blood of goats **and** calves, but **with His own blood** He entered the Most Holy Place once for all, having **obtained eternal redemption** [FOR US].* Hebrews 9:11-12

We have eternal redemption for our sins because the blood of GOD was shed to pay for our sin debt...eternally. Past, present, and future sins are covered by JESUS's blood because HE destroyed the power of the sin nature which we all inherited from Adam's fall.

NO CATHOLIC MASSES, the Eucharist, the communion, the LORD's Table allow us to ingest The LORD and have eternal life imparted to us or deposited into our human spirit core. If that is what we expect from memorializing The LORD's sacrifice for our sins then, on judgment day, we will find ourselves most miserable of all men. JESUS implemented the ceremonial ingesting of bread and wine in light of HIS crucifixion to remind us that our sins were already judged and paid for -atoned- at *The Cross.* JESUS addressed this reality proactively in the sixth chapter of John's gospel.

So Jesus said it again, "With all the earnestness I possess I tell you this: **Unless you eat the flesh of the Messiah and drink his blood, you cannot have eternal life** *within you. 54 But anyone who does eat my flesh and drink my blood has eternal life, and I will raise him at the Last Day. 55 For my flesh is the true food, and my blood is the true drink. 56 Everyone who eats my flesh and drinks my blood is in me, and I in him.* **I live by the power of the living Father** *who sent me, and in the same way those who partake of me shall live because of me! I am the true Bread from heaven; and anyone who eats this Bread shall live forever, and not die as your fathers did-though they ate bread from heaven."* John 6:53-58 TLB

It is true that we live naturally by what we eat. Spiritually, the case is exactly the same except we must speak what we eat (the Word of GOD) fully benefit from it nutritionally. Just as it is not the physical water which Jesus was offering the woman at the well; it is not the natural water of baptism which connects our spirits to the eternal Spirit of GOD in a geyser looking wormhole which Jesus spoke of in John 4:14. It is adding faith in what the Word says to the actions we take which deposit eternal life in our resurrected spirit core. It is cleansing under the fountain of Jesus' blood which does so but picturing a natural blood bath clearly leads us to a spiritual conclusion.

So if anyone eats this bread and drinks from this cup of the Lord **in an unworthy manner**, *he is guilty of sin against the body and the blood of the Lord. 28 That is why a man should examine himself carefully before eating the bread and drinking from the cup. 29 For if he eats the bread and drinks from the cup unworthily*, **not thinking about the body of Christ and what it means**, *he is eating and drinking God's judgment upon himself; for he is trifling with the death of Christ. 30 That is why many of you are weak and sick, and some have even died. 31 But if you* **carefully examine yourselves before eating you will not need to be judged and punished**. *32 Yet, when we are judged and punished by the Lord, it is so that we will not be condemned with the rest of the world.*   1 Corinthians 11:27-32 TLB

Is it possible that we carefully examine ourselves on the basis of the Levitical practice of offering sacrifice rather than on the basis of our assumptions? By that I mean that maybe we DON'T ONLY take inventory of our sins, flaws, failings, and wrongdoings and then ingest The LORD's flesh and blood following a whispered plea for forgiveness, but we also and much more importantly conduct the examination based on the Levitical model. What is that?

The person bringing the animal for sacrifice was never examined by the High Priest before the High Priest sacrificed the animal which was being presented for the sin offering. The High Priest only examined the sacrifice, and we should only examine the sacrifice which we are presenting at The LORD's Table. To not properly "discern" the LORD's body is to not add faith to our actions. Going through the motions of communion without adding faith (speaking the Word) to it is just as dangerous as treating the cracker and "juice" like a recreational snack.

You Are The Temple Of GOD
1 Corinthians 6:19

Know you not that you are the new Holy of Holies?
The presence of GOD on feet.
Mobile Tabernacle

The Lamb Of GOD Sacrificed On *The* Cross.

Even during the Levitical practice, the individual was rife with regret, haunting memories, and shame for some as their sacrifice was being evaluated by the High Priest but as long as the sacrifice qualified then the individual's sins were covered once the blood was shed and the sacrificed was cooked in a fire that burned continuously .

Our Sacrifice qualified to cover our sins forever; past, present, and future and was burned but the pangs of death and Hell could not hold on to HIM according to Acts 2:24. So, for us to carefully examine ourselves we must find our assurance and peace of mind in ingesting the flesh and blood of The LORD worthily by repeatedly realizing and even vocalizing the fact and reality that we are made right with Almighty Most High GOD based on the qualifications of our sacrifice - JESUS on the cross – and not any busy religious work we've engaged in with the subconscious motivation or intent that the work that we've done is some sort of perverted penance.

Perverted penance because the only thing which makes us right with Almighty Most High GOD is faith in what JESUS has done for and in us and nothing which we have done ourselves including confessional frequency or Eucharist ingestion.

Still, because faith in *The* Cross, the blood, and the name of JESUS opens a wormhole and possibly multiple wormholes which Heaven uses to hastily watch over the Word of GOD to ensure that it is carried out according to Jeremiah 1:12.

Dreams Are Potential Wormholes.

Potentially b/c thin line between night visions, dreams, waking dreams (trances) and eating too late before going to sleep.

NOTES

_____

_____

_____

_____

_____

_____

_____

_____

_____

_____

_____

_____

_____

_____

_____

_____

_____

_____

_____

_____

NOTES

NOTES

_____

_____

_____

_____

_____

_____

_____

_____

_____

_____

_____

_____

_____

_____

_____

_____

_____

_____

_____

_____

_____

_____

NOTES

NOTES

_____

_____

_____

_____

_____

_____

_____

_____

_____

_____

_____

_____

_____

_____

_____

_____

_____

_____

_____

NOTES

NOTES

NOTES

WORMHOLES FROM HEAVEN

---

[1]From **spaceplace.nasa.gov**/what-is-gravity
An invisible force that pulls objects toward each other. Earth's gravity is what keeps you on the ground and what makes things fall. ... So, the closer objects are to each other, the stronger their gravitational pull is. Earth's gravity comes from all its mass.

[2] **1 Corinthians 15:3-7** For I delivered to you as of first importance what I also received: that Christ died for our sins in accordance with the Scriptures, 4 that he was buried, that he was raised on the third day in accordance with the Scriptures, 5 and that he appeared to Cephas, then to the twelve. 6 Then he appeared to more than five hundred brothers at one time, most of whom are still alive, though some have fallen asleep. 7 Then he appeared to James, then to all the apostles.

[3] **Luke 24:44-49**   Then he said to them, "These are my words that I spoke to you while I was still with you, that everything written about me in the Law of Moses and the Prophets and the Psalms must be fulfilled." 45 Then he opened their minds to understand the Scriptures, 46 and said to them, "Thus it is written, that the Christ should suffer and on the third day rise from the dead, 47 and that repentance for the forgiveness of sins should be proclaimed in his name to all nations, beginning from Jerusalem. 48 You are witnesses of these things. 49 And behold, I am sending the promise of my Father upon you. But stay in the city until you are clothed with power from on high."

[4] **Mark 13:11** And when they bring you to trial and deliver you over, do not be anxious beforehand what you are to say, but say whatever is given you in that hour, for it is not you who speak, but the Holy Spirit.
Matthew 10:19-20 But when they deliver you up, do not worry about how or what you should speak. For it will be given to you in that hour what you should speak; [20] for it is not you who speak, but the Spirit of your Father who speaks in you.

[5] **John 4:13-15** Jesus answered and said to her, "Whoever drinks of this water will thirst again, 14 but whoever drinks of the water that I shall give him will never thirst. But the water that I shall give him will become in him a fountain of water springing up into everlasting life." 15 The woman said to Him, "Sir, give me this water, that I may not thirst, nor come here to draw."

[6] **John 4:16-18**   Jesus said to her, "Go, call your husband, and come here." 17 The woman answered him, "I have no husband." Jesus said to her, "You are right in saying, 'I have no husband'; 18 for you have had five husbands, and the one you now have is not your husband. What you have said is true."

www.ingramcontent.com/pod-product-compliance
Lightning Source LLC
Chambersburg PA
CBHW021406210526
45463CB00001B/245